Fixed Point Theory in Distance Spaces

William Kirk • Naseer Shahzad

Fixed Point Theory in Distance Spaces

 Springer

William Kirk
Department of Mathematics
University of Iowa
Iowa City, IA, USA

Naseer Shahzad
Department of Mathematics
King Abdulaziz University
Jeddah, Saudi Arabia

ISBN 978-3-319-36405-6   ISBN 978-3-319-10927-5 (eBook)
DOI 10.1007/978-3-319-10927-5
Springer Cham Heidelberg New York Dordrecht London

Mathematics Subject Classification (2000): 54H25, 51K10, 54C05, 47H09

Printed on acid-free paper

Springer is part of Springer Science+Business Media (www.springer.com)

ABSTRACT. Traditionally, a large body of metric fixed point theory has been couched in a functional analytic framework. This aspect of the theory has been written about extensively. This survey treats the purely metric aspects of the theory—specifically results that do not depend on any algebraic structure of the underlying space. The focus is on (I) metric spaces satisfying additional geometric conditions, (II) metric spaces with geodesic structures, and (III) semimetric spaces satisfying relaxed versions of the triangle inequality.

Preface

Mathematicians interested in topology typically give an abstract set a "topological structure" consisting of a collection of subsets of the given set to determine when points are "near" each other. People interested in geometry need a more rigid notion of nearness. This usually begins with assigning a symmetric "distance" to each two points of a set, resulting in the notion of a semimetric. With the addition of the triangle inequality, one passes to a metric space. This will be our point of departure.

There are four classical fixed point theorems against which metric extensions are usually checked. These are, respectively, the Banach contraction mapping principle, Nadler's well-known set-valued extension of that theorem, the extension of Banach's theorem to nonexpansive mappings, and Caristi's theorem. These comparisons form a significant component of this survey.

This exposition is divided into three parts. In Part I we discuss some aspects of the purely metric theory, especially Caristi's theorem and its relatives. Among other things, we discuss these theorems in the context of their logical foundations. We omit a discussion of the well-known Banach Contraction Principle and its many generalizations in Part I because this topic is well known and has been reviewed extensively elsewhere (see, e.g., [117]). In Part II we discuss classes of spaces which, in addition to having a metric structure, also have geometric structure. These specifically include the geodesic spaces, length spaces, and CAT(0) spaces. In Part III we turn to distance spaces that are not necessarily metric. These include certain distance spaces which lie strictly between the class of semimetric spaces and the class of metric spaces, as well as other spaces whose distance properties do not fully satisfy the metric axioms.

We make no attempt to explain all aspects of the topics we cover nor to present a compendium of all known facts, especially since the theory continues to expand at a rapid rate. Any attempt to provide the latest tweak on the various theorems we discuss would surely be outdated before reaching print. Our objective rather is to present a concise accessible document which can be used as an introduction to the subject and its central themes. We include proofs selectively, and from time to time we mention open problems. The material in this exposition is collected together here for the first time. Those wishing to investigate these topics deeper are referred to the original sources. We have attempted to include details in those instances where the sources are not readily available. This might be the case, for example, when the source is in a conference proceedings. Also some results appear here for the first time.

Many of the concepts introduced here have found interesting applications. Indeed some were motivated by attempts to address both mathematical and applied problems. Other concepts we discuss are more formal in nature and have yet to find any serious application; indeed some may never. However our hope is that this discussion will suggest directions for those interested in further research in this area.

The first author lectured on portions of the material covered in this monograph to students and faculty at King Abdulaziz University. He wishes to thank them for providing an attentive and critical audience. Both authors express their gratitude to Rafa Espínola for calling attention to a number of oversights in an earlier draft of this manuscript.

Iowa City, IA, USA   William Kirk
Jeddah, Saudi Arabia   Naseer Shahzad

Contents

Part I

Metric Spaces

CHAPTER 1

Introduction

At the outset we adopt the classical terminology of W.A. Wilson [216]. (The term "semimetric space" (halb-metrischer Raume) is likely due to Karl Menger [153].)

DEFINITION 1.1. Let X be a set and let $d : X \times X \to \mathbb{R}$ be a mapping satisfying for each $x, y \in X$:

  I. $d(x, y) \geq 0$, and $d(x, y) = 0 \Leftrightarrow x = y$;
  II. $d(x, y) = d(y, x)$.

Then the pair (X, d) is called a *semimetric space.*

In such a space, convergence of sequences is defined in the usual way: A sequence $\{x_n\} \subseteq X$ is said to *converge* to $x \in X$ if $\lim_{n \to \infty} d(x_n, x) = 0$. Also a sequence is said to be *Cauchy* if for each $\varepsilon > 0$ there exists $N \in \mathbb{N}$ such that $m, n \geq N \Rightarrow d(x_m, x_n) < \varepsilon$. The space (X, d) is said to be *complete* if every Cauchy sequence has a limit.

With such a broad definition of distance, three problems are immediately obvious: (i) *There is nothing to assure that limits are unique (thus the space need not be Hausdorff); (ii) a convergent sequence need not be a Cauchy sequence; (iii) the mapping $d(x, \cdot) : X \to \mathbb{R}$ need not even be continuous.* These facts preclude an effective topological theory in such a general setting.

With the introduction of the triangle inequality problems (i)–(iii) are simultaneously eliminated.

VI. (Triangle Inequality) *With X and d as in Definition 1.1 assume also that for each $x, y, z \in X$:*

$$d(x, y) \leq d(x, z) + d(z, y).$$

DEFINITION 1.2. A pair (X, d) satisfying Axioms I, II, and VI is called a *metric space.*

A metric space (X, d) is said to be *metrically convex* (or Menger convex) if given any two points $p, q \in X$ there exists a point $z \in X$, $p \neq z \neq q$, such that

$$d(p, z) + d(z, q) = d(p, q).$$

© Springer International Publishing Switzerland 2014
W. Kirk, N. Shahzad, *Fixed Point Theory in Distance Spaces,*
DOI 10.1007/978-3-319-10927-5_1

Karl Menger was a pioneer in the axiomatic study of distance spaces, and he was the first to discover the following fact.

THEOREM 1.1 ([153]). *Any two points of a complete and metrically convex metric space are the endpoints of at least one metric segment.*

See [28, p. 41] for a proof of this theorem due to N. Aronszajn. Menger based the original proof of his classical result on transfinite induction. A proof based on Caristi's theorem is given in [113].

In his study [216], Wilson introduced three axioms in addition to I and II which are weaker than VI. These are the following:

III. *For each pair of (distinct) points $x, y \in X$ there is a number $r_{x,y} > 0$ such that for every $z \in X$*

$$r_{x,y} \le d(x,z) + d(z,y).$$

IV. *For each point $x \in X$ and each $k > 0$ there is a number $r_{x,k} > 0$ such that if $y \in X$ satisfies $d(x,y) \ge k$ then for every $z \in X$*

$$r_{x,k} \le d(x,z) + d(z,y).$$

V. *For each $k > 0$ there is a number $r_k > 0$ such that if $x, y \in X$ satisfy $d(x,y) \ge k$ then for every $z \in X$*

$$r_k \le d(x,z) + d(z,y).$$

Obviously if Axiom V is strengthened to $r_k = k$, then the space becomes metric. W.A. Wilson asserts in [216] that E.W. Chittenden [53] has shown (using an equivalent definition) that a semimetric space satisfying Axiom V is always metrizable. (We have not independently verified this assertion.)

Axiom III in a semimetric space (X, d) is equivalent to the assertion that there do not exist distinct points $x, y \in X$ and a sequence $\{z_n\} \subseteq X$ such that $d(x, z_n) + d(y, z_n) \to 0$ as $n \to \infty$. Thus, as Wilson observes, the following is self-evident.

PROPOSITION 1.1. *In a semimetric space Axiom* III *implies that limits are unique.*

For $r > 0$, let $U(p; r) = \{x \in X : d(x, p) < r\}$. Then Axiom III is also equivalent to the assertion that X is Hausdorff in the sense that given any two distinct points $x, y \in X$ there exist positive numbers r_x and r_y such that $U(x; r_x) \cap U(y; r_y) = \emptyset$.

DEFINITION 1.3. Let (X, d) be a semimetric space. Then the distance function d is said to be *continuous* if for any sequences $\{p_n\}, \{q_n\} \subseteq X$, $\lim_{n\to\infty} d(p_n, p) = 0$ and $\lim_{n\to\infty} d(q_n, q) = 0 \Rightarrow \lim_{n\to\infty} d(p_n, q_n) = d(p, q)$.

A point p in a semimetric space X is said to be an *accumulation point* of a subset E of X if given any $\varepsilon > 0$, $U(p; \varepsilon) \cap E \neq \emptyset$. A subset of a semimetric space is said to be *closed* if it contains each of its accumulation points. A subset of a semimetric space is said to be *open* if its complement is closed. With these definitions, if X is a semimetric space with continuous distance function, then $U(p; r)$ is an open set for each $p \in X$ and $r > 0$ and moreover, X is a Hausdorff topological space [**28**, p. 11].

REMARK 1.1. *Some authors call a space satisfying Axioms I and II a symmetric space, and reserve the term semimetric space for symmetric spaces with continuous distance function. We use the classical definition in this monograph.*

CHAPTER 2

Caristi's Theorem and Extensions

2.1. Introduction

Much of the material immediately following is taken from [115]. We begin with two "equivalent" facts. The first is a well-known variational principle due to Ekeland [70, 71] and the second is the well-known Caristi Theorem [49]. Throughout we use $\mathbb{R}$ to denote the set of real numbers, $\mathbb{N}$ to denote the set of natural numbers, and $\mathbb{R}^+ = [0, \infty)$. Recall that if X is a metric space, a mapping $\varphi : X \to \mathbb{R}^+$ is said to be (sequentially) *lower semi-continuous* (l.s.c.) if given any sequence $\{x_n\}$ in X, the conditions $x_n \to x$ and $\varphi(x_n) \to r$ imply $\varphi(x) \leq r$.

THEOREM 2.1 (E). (*Ekeland [70]*) *Let* (X, d) *be a complete metric space and* $\varphi : X \to \mathbb{R}^+$ *l.s.c. Define a partial order* $\leq$ *on* X *as follows:*

$$(2.1) \qquad x \leq y \Leftrightarrow d(x, y) \leq \varphi(x) - \varphi(y), \quad x, y \in X.$$

Then $(X, \leq)$ *has a maximal element.*

THEOREM 2.2 (C). (*Caristi [49]*) *Let* X *and* φ *be as above. Suppose* $f : X \to X$ *satisfies*

$$(2.2) \qquad d(x, f(x)) \leq \varphi(x) - \varphi(f(x)), \quad x \in X.$$

Then f *has a fixed point.*

(E) $\Rightarrow$ (C).

PROOF. With X, φ as above and f as in (C), define the relation $\leq$ on X by setting

$$x \leq y \Longleftrightarrow d(x, y) \leq \varphi(x) - \varphi(y), \quad x, y \in X.$$

By (E) $(X, \leq)$ has a maximal element x^*. However by (2.2)

$$d(x^*, f(x^*)) \leq \varphi(x^*) - \varphi(f(x^*)),$$

and this in turn implies $x^* \leq f(x^*)$ so by maximality of x^* it must be the case that $f(x^*) = x^*$. $\qquad \square$

(C) $\Rightarrow$ (E).

© Springer International Publishing Switzerland 2014
W. Kirk, N. Shahzad, *Fixed Point Theory in Distance Spaces*,
DOI 10.1007/978-3-319-10927-5_2

PROOF. Assume X, φ, and $\leq$ are as in (E), and assume $(X, \leq)$ does not have a maximal element. Then for each $x \in X$ there exists $y_x \in X$ such that $x < y_x$. Define $f : X \to X$ by setting $f(x) = y_x$. Then by (2.1) $d(x, y_x) \leq \varphi(x) - \varphi(y_x)$; hence

$$d(x, f(x)) \leq \varphi(x) - \varphi(f(x)), \quad x \in X.$$

By (C) f has a fixed point x^*. But by assumption $x^* < f(x^*)$, which is a contradiction.   □

Thus it is easy to see that (E) $\Leftrightarrow$ (C). However to a logician these two results are not equivalent. In particular the implication (C) $\Rightarrow$ (E) invokes the Axiom of Choice (AC). In fact, N. Brunner [42] has shown that *any* proof of (E) requires at least the basic axioms of Zermelo–Fraenkel (ZF) plus a form of the Axiom of Choice called the Axiom of Dependent Choice (DC), whereas R. Mańka [143] has shown that (C) holds within (ZF). So from a purely logical point of view the two theorems are not equivalent. (DC) is strictly weaker than (AC) but strictly stronger than the Axiom of Countable Choice.

Brézis and Browder derive Ekeland's Theorem from an order principle (see Theorem 2.4 below) which requires only ZFDC. They then derive Caristi's Theorem as in the implication (E) $\Rightarrow$ (C) above. Hence Choice is invoked at this step. However in [87] it is shown that Caristi's theorem can be derived directly from the order principle of Brézis and Browder without recourse to Ekeland's Theorem. We give a similar proof below (see Theorem 2.3).

In the chart below we list the authors of some of the early proofs of Caristi's theorem, the methods, and the axioms used. See Sect. 13.2 for a quick proof using Zorn's Lemma.

Author	Method	Axioms
Caristi (1976) [49]	Transfinite Induction	ZFAC
Wong (1976) [217]	Transfinite Induction	ZFAC
Kirk (1976) [113]	Zorn's Lemma	ZFAC
Brøndsted (1976) [38]	Zorn's Lemma	ZFAC
Browder (1976) [39]	Mathematical Induction	ZFDC
Brézis–Browder (1976) [34]	Mathematical Induction	ZFDC
Penot (1976) [169]		ZFDC
Siegel (1977) [202]		ZFDC
Pasicki (1978) [167]		ZFAC
Mańka (1988) [144]		ZF
Goebel–Kirk (1990) [87]		ZFDC

It is interesting that to this day Caristi's Theorem continues to be "generalized" (see, e.g., [32, 206]). Indeed Caristi's name appears in the *titles* of over one hundred papers. It would be a huge undertaking to see how many of the literally dozens of generalizations and/or extensions of Caristi's Theorem can be obtained without at least assuming DC. At the same time many "extensions" of Caristi's theorem turn out to be consequences of Caristi's theorem. The next section provides an illustration of this very fact.

2.2. A Proof of Caristi's Theorem

The paper [32] uses as its point of departure the following definition introduced in Kirk and Saliga [126]. (The idea has also been credited to [52]. However the talk in which this definition was introduced, and on which [126] is based, was delivered at the meeting of the World Congress of Nonlinear Analysts in Catania, Sicily, July, 2000.)

DEFINITION 2.1. Let X be a metric space. A mapping $\varphi : X \to \mathbb{R}$ is said to be [*sequentially*] *lower semicontinuous from above* (l.s.c.a.) if given any net [*sequence*] $\{x_\alpha\}$ in X, whenever $x_\alpha \to x$ and $\{\varphi(x_\alpha)\} \to r$ *is* nonincreasing $(\varphi(x_\alpha) \searrow r)$, then $\varphi(x) \leq r$.

It is shown in [126] that this weaker lower semicontinuity suffices for Caristi's Theorem, a fact which leads directly to another proof of the Downing–Kirk [63] extension of Caristi's Theorem.

THEOREM 2.3 ([126]). *Suppose (X, d) is complete, suppose $\varphi : X \to \mathbb{R}$ is bounded below and lower semicontinuous from above, and suppose $f : X \to X$ is an arbitrary mapping satisfying*

$$(2.3) \qquad d(x, f(x)) \leq \varphi(x) - \varphi(f(x))$$

for all $x \in X$. Then f has a fixed point.

We shall derive this theorem from the following order principle due to Brézis and Browder [34].

THEOREM 2.4 (Brézis–Browder Order Principle). *Let $(X, \preceq)$ be a partially ordered set, and for $x \in X$, set $S(x) = \{y \in X : x \preceq y\}$. Suppose $\psi : X \to \mathbb{R}$ satisfies:*

(a) *$x \preceq y$ and $x \neq y \Rightarrow \psi(x) < \psi(y)$;*
(b) *for any increasing sequence $\{x_n\}$ in X such that $\psi(x_n) \leq C < \infty$ for all n there exists some $y \in X$ such that $x_n \preceq y$ for all n;*
(c) *for each $x \in X$, $\psi(S(x))$ is bounded above.*

Then for each $x \in X$ there exists $x^* \in S(x)$ such that x^* is maximal in $(X, \preceq)$, that is, $S(x^*) = \{x^*\}$.

PROOF OF THEOREM 2.3. Let $\preceq$ denote the Brøndsted order in X. Thus for $x, y \in X$, $x \preceq y \Leftrightarrow d(x, y) \leq \varphi(x) - \varphi(y)$. Now let $\psi = -\varphi$. Then condition (a) of Theorem 2.4 is obvious, and condition (c) follows from the fact that φ is bounded below. To see that (b) holds, suppose $\{x_n\}$ is an increasing sequence in $(X, \preceq)$ which satisfies $\psi(x_n) \le C < \infty$ for all $n \in \mathbb{N}$. Then $\{\varphi(x_n)\}$ is a decreasing sequence in $\mathbb{R}$, so there exists $r \in \mathbb{R}$ such that $\lim_{n \to \infty} \varphi(x_n) = r$. Since $\{\varphi(x_n)\}$ is decreasing, for any $m > n$,

$$\lim_{m,n \to \infty} d(x_n, x_m) \le \lim_{m,n \to \infty} [\varphi(x_n) - \varphi(x_m)] = 0.$$

Therefore $\{x_n\}$ is a Cauchy sequence in X. Hence there exists $x \in X$ such that $\lim_{n \to \infty} x_n = x$. Since $\varphi(x_n) \searrow r$, $\varphi(x) \le r$ and it follows that

$$\begin{aligned} d(x_n, x) &\le \lim_{m \to \infty} d(x_n, x_m) \le \lim_{m \to \infty} [\varphi(x_n) - \varphi(x_m)] \\ &= \varphi(x_n) - r \le \varphi(x_n) - \varphi(x). \end{aligned}$$

Therefore x is an upper bound for $\{x_n\}$ in $(X, \preceq)$, proving (b) of Theorem 2.4. By Theorem 2.4 $(X, \preceq)$ has a maximal element, say x^*. Since condition (2.3) implies $x^* \preceq f(x^*)$ it must be the case that $f(x^*) = x^*$.   □

Theorem 2.3 contains the following theorem due to Downing and Kirk [**63**].

THEOREM 2.5. *Suppose (X, d) and (Y, ρ) are complete metric spaces, let $f : X \to Y$ be a closed mapping, and let $\phi : X \to \mathbb{R}$ be lower semicontinuous and bounded below. Let $g : X \to X$ satisfy*

$$\max\{d(x, g(x)), c\rho(f(x), f(g(x)))\} \le \phi(f(x)) - \phi(f(g(x)))$$

for some constant $c > 0$ and all $x \in X$. Then g has a fixed point.

PROOF. Introduce the metric D on X by setting

$$D(x, y) = \max\{d(x, y), c\rho(f(x), f(y))\}$$

for all $x, y \in X$. It is easy to check that (X, D) is a complete metric space. Now let $\varphi := \phi \circ f$, and define

$$x \preceq y \Leftrightarrow D(x, y) \le \varphi(x) - \varphi(y)$$

for $x, y \in X$. Now suppose $\{x_n\}$ is decreasing in $(X, \preceq)$. Then $\{\varphi(x_n)\}$ is decreasing, so there exists $r \in \mathbb{R}$ such that $\lim_{n \to \infty} \varphi(x_n) = r$. Also

$$\lim_{m,n \to \infty} D(x_n, x_m) = \lim_{m,n \to \infty} \max\{d(x_n, x_m), c\rho(f(x_n), f(x_m))\} = 0,$$

and this implies that both $\{x_n\}$ and $\{f(x_n)\}$ are Cauchy sequences in (X, d) and (Y, ρ), respectively. Hence there exist $x \in X$, $y \in Y$ such that $\lim_{n \to \infty} x_n = x$ and $\lim_{n \to \infty} f(x_n) = y$. Since f is a closed mapping, $f(x) = y$. Also, since ϕ is lower semicontinuous we have

$$\varphi(x) = \phi(y) \le \lim_{n \to \infty} \phi \circ f(x_n) = \lim_{n \to \infty} \varphi(x_n).$$

This shows that φ is lower semicontinuous from above. Therefore Theorem 2.3 can be applied directly to the complete metric space (X, D). Since $D(x, g(x)) \le \varphi(x) - \varphi(g(x))$, we conclude g has a fixed point.   □

2.3. Suzuki's Extension

We now turn to the main result of the paper [**206**]. Suzuki shows that results of [**15, 16**] follow directly from his result.

THEOREM 2.6. *Let (X, d) be a complete metric space. Let $f : X \to X$, and let $\varphi : X \to \mathbb{R}^+$ be lower semicontinuous. Let $\Psi : X \to \mathbb{R}^+$ satisfy*

$$\sup \left\{ \Psi(x) : x \in X, \ \varphi(x) \leq \inf_{w \in X} \varphi(w) + \eta \right\} < \infty$$

for some $\eta > 0$. Assume that

$$d(x, f(x)) \leq \Psi(x)(\varphi(x) - \varphi(f(x)))$$

for all $x \in X$. Then f has a fixed point.

PROOF. When $\Psi(x) > 0$ then $\varphi(f(x)) \leq \varphi(x)$ by assumption, and when $\Psi(x) = 0$, $x = f(x)$, so $\varphi(f(x)) \leq \varphi(x)$ for all $x \in X$. Set

$$Y = \left\{ x \in X : \varphi(x) \leq \inf_{w \in X} \varphi(w) + \eta \right\} \text{ and } \gamma = \sup_{w \in Y} \Psi(w) < \infty.$$

Note that Y is closed and hence complete because X is complete and φ is lower semicontinuous. It is clear that $Y \neq \emptyset$, and because $\varphi(f(x)) \leq \varphi(x)$, $f(Y) \subseteq Y$. Also

$$d(x, f(x)) \leq \Psi(x)(\varphi(x) - \varphi(f(x))) \leq \gamma(\varphi(x) - \varphi(f(x)))$$

for all $x \in Y$. Since $x \longmapsto \gamma\varphi(x)$ is lower semicontinuous, f has a fixed point in Y by Caristi's Theorem. $\qquad\square$

REMARK 2.1. *In order to apply Caristi's Theorem, it suffices only to know that Ψ is lower semicontinuous from above. However this assumption is not enough to assure that Y is complete.*

2.4. Khamsi's Extension

In [**50**] Caristi posed the following problem (which he attributed to one of the present writers). Does Theorem 2.2 remain true if instead of (2.2) it is merely assumed that for some $p > 1$,

$$(d(x, f(x)))^p \leq \varphi(x) - \varphi(f(x)), \quad x \in X?$$

Some time ago it was shown by Bae and Park [**14**] that the answer is negative. More recently Khamsi [**106**] has given another negative answer to this question.

Example. ([106]) Let $p > 1$, let $x_n := \sum_{i=1}^{n} \frac{1}{i}$, and let $X = \{x_n : n \in \mathbb{N}\}$. Then X is a closed (discrete) subset of $\mathbb{R}^+$ and is therefore complete. (If $m > n$, then $d(x_n, x_m) \geq \frac{1}{m}$.) Define $f : X \to X$ by taking $f(x_n) = x_{n+1}$ for all $n \geq 1$. Now define $\varphi(x_n) = \sum_{i=n+1}^{\infty} \frac{1}{i^p}$. Then

$$
\begin{aligned}
(d(x_n, f(x_n)))^p &= \frac{1}{(n+1)^p} \\
&= \sum_{i=n+1}^{\infty} \frac{1}{i^p} - \sum_{i=n+2}^{\infty} \frac{1}{i^p} \\
&= \varphi(x_n) - \varphi(x_{n+1}) \\
&= \varphi(x_n) - \varphi(f(x_n)).
\end{aligned}
$$

Clearly f is fixed point free. Also note that φ is continuous (because X is discrete) and f is even nonexpansive.

Khamsi then turns to the question of whether there exist positive functions $\eta : \mathbb{R}^+ \to \mathbb{R}^+$ with the property that if $f : X \to X$ (X complete) satisfies

$$\eta(d(x, f(x))) \leq \varphi(x) - \varphi(f(x)), \qquad x \in X,$$

for some lower semicontinuous $\varphi : X \to \mathbb{R}^+$, then f has a fixed point. He gives an affirmative answer in the form of the following theorem.

The standing assumptions are these: $\eta : \mathbb{R}^+ \to \mathbb{R}^+$ is nondecreasing, continuous, and such that there exists $c > 0$ and $\delta_0 > 0$ such that for any $t \in [0, \delta_0]$, $\eta(t) \geq ct$. Because η is continuous, there exists $\varepsilon_0 > 0$ such that $\eta^{-1}([0, \varepsilon_0]) \subset [0, \delta_0]$.

THEOREM 2.7. *Suppose X is a complete metric space and $\varphi : X \to \mathbb{R}^+$ lower semicontinuous. Define the relation $\prec$ on X by setting*

$$x \prec y \Leftrightarrow \eta(d(x, y)) \leq \varphi(y) - \varphi(x), \qquad x, y \in X,$$

where η is as above. Then $(X, \prec)$ has a minimal element x^* (i.e., $x \prec x^* \Rightarrow x = x^*$).

PROOF. ([106]) Set $\varphi_0 = \inf\{\varphi(x) : x \in X\}$. For any $\varepsilon > 0$, set

$$X_\varepsilon = \{x \in X : \varphi(x) \leq \varphi_0 + \varepsilon\}.$$

Since φ is lower semicontinuous, X_ε is a closed nonempty subset of X. (This uses the fact that φ is lower semicontinuous. Suppose $\{x_n\} \subset X_\varepsilon$ and $x_n \to x$. Then $\varphi(x_n) \leq \varphi_0 + \varepsilon$, so $\varphi(x) \leq \varphi_0 + \varepsilon$ i.e., $x \in X_\varepsilon$.) Also, if $x, y \in X_\varepsilon$ and if $x \prec y$, then

$$\eta(d(x, y)) \leq \varphi(y) - \varphi(x)$$

which implies

$$\varphi_0 \leq \varphi(x) \leq \varphi(y) \leq \varphi_0 + \varepsilon.$$

Hence $\eta \left(d \left(x, y \right) \right) \leq \varepsilon$. In particular, if $x, y \in X_{\varepsilon_0}$ and $x \prec y$, then

$$c \left(d \left(x, y \right) \right) \leq \eta \left(d \left(x, y \right) \right) \leq \varphi \left(y \right) - \varphi \left(x \right).$$

Now on X_{ε_0} define the new relation $\prec_*$ by

$$x \prec_* y \Leftrightarrow d \left(x, y \right) \leq \frac{1}{c} \varphi \left(y \right) - \frac{1}{c} \varphi \left(x \right), \qquad x, y \in X_{\varepsilon_0}.$$

Clearly $\left(X_{\varepsilon_0}, \prec_* \right)$ is a partial order with all the necessary assumptions for securing, via Zorn's Lemma, an element $x^* \in X_{\varepsilon_0}$ which is minimal relative to $\prec_*$.

Now let $x \in X$ satisfy $x \prec x^*$. Then $\eta \left(d \left(x, x^* \right) \right) \leq \varphi \left(x^* \right) - \varphi \left(x \right)$, so $\varphi \left(x \right) \leq \varphi \left(x^* \right) \leq \varphi_0 + \varepsilon_0$, i.e., $x \in X_{\varepsilon_0}$. As before, $\eta \left(d \left(x, x^* \right) \right) \leq \varepsilon_0$ and this implies

$$cd \left(x, x^* \right) \leq \eta \left(d \left(x, x^* \right) \right) \leq \varphi \left(x^* \right) - \varphi \left(x \right).$$

Since x^* is minimal in $\left(X_{\varepsilon_0}, \prec_* \right)$ we have $x = x^*$. $\qquad\square$

The following is Theorem 3 of [**106**].

THEOREM 2.8. *Let X be a complete metric space and let $f : X \to X$ be a mapping such that for all $x \in X$*

$$\eta \left(d \left(x, f \left(x \right) \right) \right) \leq \varphi \left(x \right) - \varphi \left(f \left(x \right) \right),$$

where η and φ are as in Theorem 2.7. Then f has a fixed point.

PROOF. Define the relation $\prec$ as in Theorem 2.7. Obviously $f \left(x \right) \prec x$ for any $x \in X$. In particular, if x^* is a minimal element in $\left(X, \prec \right)$, it must be the case that $f \left(x^* \right) = x^*$. $\qquad\square$

We now turn to a variant of Khamsi's Theorem.

THEOREM 2.9. *Suppose X is a complete metric space and suppose $\varphi : X \to \mathbb{R}^+$ is bounded below and sequentially lower semicontinuous from above. Define the relation $\prec$ on X by setting*

$$x \prec y \Leftrightarrow \eta \left(d \left(x, y \right) \right) \leq \varphi \left(x \right) - \varphi \left(y \right), \qquad x, y \in X,$$

where η is as in Theorem 2.7. Then $\left(X, \prec \right)$ has a maximal element x^* (i.e., $x^* \prec x$ for $x \in X \Rightarrow x = x^*$).

PROOF. Set $\varphi_0 = \inf \left\{ \varphi \left(x \right) : x \in X \right\}$. For any $\varepsilon > 0$, set

$$X_\varepsilon = \left\{ x \in X : \varphi \left(x \right) \leq \varphi_0 + \varepsilon \right\}.$$

If $x, y \in X_\varepsilon$ and if $x \prec y$, then

$$\eta \left(d \left(x, y \right) \right) \leq \varphi \left(x \right) - \varphi \left(y \right)$$

which implies

$$\varphi_0 \leq \varphi \left(y \right) \leq \varphi \left(x \right) \leq \varphi_0 + \varepsilon.$$

Hence $\eta \left(d \left(x, y \right) \right) \leq \varepsilon$. In particular, if $x, y \in X_{\varepsilon_0}$ and $x \prec y$, then

$$cd \left(x, y \right) \leq \eta \left(d \left(x, y \right) \right) \leq \varphi \left(x \right) - \varphi \left(y \right).$$

Now on X_{ε_0} we define the new relation $\prec_*$ by

$$x \prec_* y \Leftrightarrow d(x,y) \leq \frac{1}{c}\varphi(x) - \frac{1}{c}\varphi(y), \qquad x,y \in X_{\varepsilon_0}.$$

Set $\psi := -\frac{1}{c}\varphi$ and define $x \leq y \Leftrightarrow d(x,y) \leq \psi(y) - \psi(x)$. We now show that $(X_{\varepsilon_0}, \leq)$ has a maximal element. (Notice that we are not assuming X_{ε_0} is complete.) Condition (a) of Theorem 2.4 is obvious, and condition (c) follows from the fact that φ is bounded below. To see that (b) holds, suppose $\{x_n\}$ is an increasing sequence in $(X_{\varepsilon_0}, \leq)$ which satisfies $\psi(x_n) \leq C < \infty$ for all n. Then $\{\varphi(x_n)\}$ is a decreasing sequence in $\mathbb{R}$, so there exists $r \in \mathbb{R}$ such that $\lim_{n \to \infty} \varphi(x_n) = r$. Since for any $m > n$,

$$\lim_{m,n \to \infty} cd(x_n, x_m) \leq \lim_{m,n \to \infty} [\varphi(x_n) - \varphi(x_m)] = 0.$$

It follows that $\{x_n\}$ is a Cauchy sequence in X, and since X is complete there exists $x \in X$ such that $\lim_n x_n = x$. Since $\varphi(x_n) \searrow r$ and φ is lower semicontinuous from above, $\varphi(x) \leq r$. However $x_n \in X_{\varepsilon_0} \Rightarrow \varphi(x_n) \leq \varphi_0 + \varepsilon_0$. Therefore $r \leq \varphi_0 + \varepsilon$; hence $\varphi(x) \leq \varphi_0 + \varepsilon_0$, and so $x \in X_{\varepsilon_0}$. It follows that

$$\begin{aligned} cd(x_n, x) &= \lim_{m \to \infty} cd(x_n, x_m) \leq \lim_{m \to \infty} [\varphi(x_n) - \varphi(x_m)] \\ &= \varphi(x_n) - r \leq \varphi(x_n) - \varphi(x). \end{aligned}$$

Thus x is an upper bound for $\{x_n\}$ in $(X_{\varepsilon_0}, \leq)$. By Theorem 2.4, there exists a maximal element x^* in $(X_{\varepsilon_0}, \leq)$, and in turn x^* is a maximal element in $(X_{\varepsilon_0}, \prec_*)$.

Now let $x \in X$ satisfy $x^* \prec x$. Then $\eta(d(x,x^*)) \leq \varphi(x^*) - \varphi(x)$, so $\varphi(x) \leq \varphi(x^*) \leq \varphi_0 + \varepsilon_0$, i.e., $x \in X_{\varepsilon_0}$. As before, $\eta(d(x,x^*)) \leq \varepsilon_0$ and this implies

$$cd(x,x^*) \leq \eta(d(x,x^*)) \leq \varphi(x^*) - \varphi(x).$$

Since x^* is maximal in $(X_{\varepsilon_0}, \prec_*)$ we have $x = x^*$. $\qquad\square$

Since the Brézis–Browder order principle does not require Zorn's Lemma, the preceding result yields a more "constructive" proof of a slight generalization of Khamsi's Theorem.

COROLLARY 2.1. *Suppose X is a complete metric space and suppose $\varphi : X \to \mathbb{R}$ is bounded below and sequentially lower semicontinuous from above. Define the relation $\prec$ on X by setting*

$$x \prec y \Leftrightarrow \eta(d(x,y)) \leq \varphi(y) - \varphi(x), \qquad x,y \in X,$$

where η is as above. Then $(X, \prec)$ has a minimal element x^*.

PROOF. An element $x^* \in X$ is *maximal* in X relative to the relation

$$x \prec y \Leftrightarrow \eta(d(x,y)) \leq \varphi(x) - \varphi(y), \qquad x,y \in X,$$

if and only if x^* is *minimal* in X relative to the relation

$$x \prec y \Leftrightarrow \eta(d(x,y)) \leq \varphi(y) - \varphi(x), \qquad x,y \in X.$$

$\qquad\square$

THEOREM 2.10. *Let (X, d) be a complete metric space. Let $f : X \to X$, and let $\varphi : X \to \mathbb{R}^+$ be lower semicontinuous from above. Let $\Psi : X \to \mathbb{R}^+$ satisfy*

$$\sup \left\{ \Psi(x) : x \in X, \ \varphi(x) \leq \inf_{w \in X} \varphi(w) + \varepsilon \right\} < \infty$$

for some $\varepsilon > 0$. Introduce the relation $\prec$ on X by setting

$$x \prec y \Leftrightarrow \eta(d(x, y)) \leq \Psi(x)(\varphi(x) - \varphi(y))$$

for all $x, y \in X$. Then there is an element $x^* \in X$ that is maximal relative to $\prec$.

PROOF. First notice that $x \prec y \Rightarrow \varphi(y) \leq \varphi(x)$ for all $x, y \in X$. Let $\varphi_0 = \inf_{w \in X} \varphi(w)$, and set

$$Y = \{ x \in X : \varphi(x) \leq \varphi_0 + \varepsilon \} \text{ and } \gamma = \sup_{w \in Y} \Psi(w) < \infty.$$

Now let

$$X_\varepsilon = \left\{ x \in X : \varphi(x) \leq \varphi_0 + \gamma^{-1} \varepsilon \right\}$$

and introduce the relation $\prec_*$ on X_ε by setting

$$x \prec_* y \Leftrightarrow \eta(d(x, y)) \leq \Psi(x)(\varphi(x) - \varphi(y)).$$

which in turn implies

$$\varphi_0 \leq \varphi(y) \leq \varphi(x) \leq \varphi_0 + \gamma^{-1} \varepsilon.$$

In particular $\varphi(x) - \varphi(y) \leq \gamma^{-1} \varepsilon$. Let $\varepsilon_0' = \min \{ \varepsilon, \varepsilon_0 \}$. Thus if $x, y \in X_{\varepsilon_0'}$, $\eta(d(x, y)) \leq \Psi(x)(\varphi(x) - \varphi(y)) \leq \gamma(\gamma^{-1}) \varepsilon_0' \leq \varepsilon_0$. In particular

$$cd(x, y) \leq \eta(d(x, y)) \leq \Psi(x)(\varphi(x) - \varphi(y)) \leq \gamma(\varphi(x) - \varphi(y)).$$

Now let $\phi := -\dfrac{\gamma}{c} \varphi$ and introduce the new partial order $\leq$ on X_{ε_0} by setting

$$x \leq y \Leftrightarrow d(x, y) \leq \phi(y) - \phi(x).$$

It is now possible to complete the proof exactly as in the proof of Theorem 2.9.

$\square$

Observe that by taking Ψ to be the identity mapping one recovers Khamsi's theorem.

COROLLARY 2.2. *Let (X, d) be a complete metric space. Let $f : X \to X$, and let $\varphi : X \to \mathbb{R}^+$ be lower semicontinuous from above. Let $\Psi : X \to \mathbb{R}^+$ satisfy*

$$\sup \left\{ \Psi(x) : x \in X, \ \varphi(x) \leq \inf_{w \in X} \varphi(w) + \varepsilon \right\} < \infty$$

for some $\varepsilon > 0$. Assume that

$$\eta(d(x, f(x))) \leq \Psi(x)(\varphi(x) - \varphi(f(x)))$$

for all $x \in X$. Then f has a fixed point.

PROOF. Introduce the relation $\prec$ on X by setting

$$x \prec y \Leftrightarrow \eta(d(x,y)) \leq \Psi(x)(\varphi(x) - \varphi(y))$$

for all $x, y \in X$. By Theorem 2.10, there is a point $x^* \in X$ that is maximal relative to this relation. However by assumption, $x^* \prec f(x^*)$. It follows that $f(x^*) = x^*$. $\qquad\qquad\square$

2.5. Results of Z. Li

In [139] Z. Li shows that one can actually derive Khamsi's results from Caristi's Theorem without assumptions on the continuity and the subadditivity of η. We summarize Li's results of [139] here. Throughout, (X, d) denotes a complete metric space. A mapping $f : X \to X$ is said to be a *Caristi type mapping* if

(2.4) $\qquad\qquad \eta(d(x, f(x))) \leq \varphi(x) - \varphi(f(x)) \qquad \forall x \in X,$

where $\eta : \mathbb{R}^+ \to \mathbb{R}$ and $\varphi : X \to \mathbb{R}$.

PROPOSITION 2.1. *Suppose that* $\eta : \mathbb{R}^+ \to \mathbb{R}^+$ *and the Caristi type mapping has a fixed point in* X. *Then* $\eta(0) = 0$.

PROOF. Suppose $f(x^*) = x^*$. If $\eta(0) \neq 0$, then $\eta(0) > 0$. Hence from (2.4)

(2.5) $\qquad 0 < \eta(0) = \eta(d(x^*, f(x^*))) \leq \varphi(x^*) - \varphi(f(x^*)) = 0$

which is a contradiction. Therefore $\eta(0) = 0$. $\qquad\qquad\square$

From this result it is easy to see that Khamsi's theorem holds if $\eta(0) = 0$. The following theorem actually reduces to an application of Caristi's theorem. This in turn shows that Khamsi's theorem is actually a consequence of Caristi's theorem.

THEOREM 2.11. *Suppose that* $\eta : \mathbb{R}^+ \to \mathbb{R}$ *with* $\eta(0) = 0$, *suppose* $\varphi : X \to \mathbb{R}$ *is lower semicontinuous, and suppose there exist* $x_0 \in X$ *and two real numbers* $a, \beta \in \mathbb{R}$ *such that*

(2.6) $\qquad\qquad \varphi(x) \geq ad(x, x_0) + \beta \qquad \forall x \in X.$

Suppose also that one of the following conditions is satisfied.

    (i) $a \geq 0$, η *is nonnegative and nondecreasing on* $W = \{d(x, y) : x, y \in X\}$, *and there exists* $c > 0$ *and* $\varepsilon > 0$ *such that*

(2.7) $\qquad\qquad \eta(t) \geq ct \qquad \forall t \in \{t \geq 0 : \eta(t) \leq \varepsilon\} \cap W;$

    (ii) $a < 0$, $\eta(t) + at$ *is nonnegative and nondecreasing on* W, *and there exist* $c > 0$ *and* $\varepsilon > 0$ *such that*

(2.8) $\qquad \eta(t) + at \geq ct \qquad \forall t \in \{t \geq 0 : \eta(t) + at \leq \varepsilon\} \cap W.$

    Then each Caristi type mapping has a fixed point in X.

PROOF. Case (i). From $a \geq 0$ and (2.6) we see that φ is bounded below on X. Let

$$(2.9) \qquad \alpha = \inf \{\varphi(x) : x \in X\}.$$

Let

$$X_\varepsilon = \{x \in X : \varphi(x) \leq \alpha + \varepsilon\}.$$

From the lower semicontinuity of φ, it follows that the set X_ε is a nonempty closed subset of X. Hence (X_ε, d) is a complete metric space. We show that $f : X_\varepsilon \to X_\varepsilon$. Since $\eta(t) \geq 0$ for each $t \in W$ and $d(x, f(x)) \in W$ for each $x \in X$ we have

$$(2.10) \qquad 0 \leq \eta(d(x, f(x))) \leq \varphi(x) - \varphi(f(x)) \qquad \forall x \in X_\varepsilon.$$

Therefore

$$(2.11) \qquad \alpha \leq \varphi(f(x)) \leq \varphi(x) \leq \alpha + \varepsilon \qquad \forall x \in X_\varepsilon.$$

This proves that $f : X_\varepsilon \to X_\varepsilon$.

For each $x \in X_\varepsilon$ we have from (2.10) and (2.11)

$$(2.12) \qquad 0 \leq \eta(d(x, f(x))) \leq \varphi(x) - \varphi(f(x)) \leq \varphi(x) - \alpha \leq \varepsilon.$$

From (2.7) and (2.12)

$$cd(x, f(x)) \leq \eta(d(x, f(x))) \leq \varphi(x) - \varphi(f(x)) \qquad \forall x \in X_\varepsilon.$$

Letting $\phi = \dfrac{1}{c}\varphi$, we now have

$$d(x, f(x)) \leq \phi(x) - \phi(f(x)) \qquad \forall x \in X_\varepsilon.$$

Therefore by Caristi's theorem, f has a fixed point in X_ε.

Case (ii). Let

$$(2.13) \qquad \psi(x) = \varphi(x) - ad(x, x_0) \quad \text{for each } x \in X.$$

From (2.6) and (2.13) it is easy to see that $\psi : X \to [\beta, \infty)$ is lower semicontinuous and bounded below on X. Let

$$(2.14) \qquad \eta_1(t) = \eta(t) + at \quad \text{for each } t \in \mathbb{R}^+.$$

Then η_1 is nonnegative and nondecreasing on W, so from (2.8) we have

$$\eta_1(t) \geq ct \quad \text{for each } t \in \{t \geq 0 : \eta_1(t) \leq \varepsilon\} \cap W.$$

On the other hand, from (2.4) and (2.13)–(2.14),

$$\begin{aligned} \eta_1(d(x, f(x))) &= \eta(d(x, f(x))) + ad(x, f(x)) \\ &\leq \varphi(x) - ad(x, x_0) - \varphi(f(x)) + ad(f(x), x_0) \\ &\leq \psi(x) - \psi(f(x)). \end{aligned}$$

The above fact and Case (i) imply that f has a fixed point. $\qquad\square$

2.6. A Theorem of Zhang and Jiang

Let $\gamma : \mathbb{R}^+ \to \mathbb{R}^+$ be a subadditive (i.e., $\gamma(t+s) \leq \gamma(t) + \gamma(s)$ for $s, t \in \mathbb{R}^+$) and increasing continuous mapping such that $\gamma^{-1}(\{0\}) = 0$. For example, $\gamma(t) = t^p$ ($0 < p \leq 1$) for $t \in \mathbb{R}^+$). Let Γ denote the collection of all such functions γ.

Let $\mathcal{A}$ denote the class of all maps $\eta : \mathbb{R}^+ \to \mathbb{R}^+$ for which there exists $\bar{\varepsilon} > 0$ and $\gamma \in \Gamma$ such that if $\eta(t) \leq \bar{\varepsilon}$, then $\eta(t) \geq \gamma(t)$.

Let $F : \mathbb{R} \to \mathbb{R}$ be an increasing, upper semi-continuous mapping such that $F(0) = 0$, $F^{-1}(\mathbb{R}^+) \subset \mathbb{R}^+$ and such that $F(t) + F(s) \leq F(t+s)$ for $t, s \geq 0$. For example,

$$F(t) = \begin{cases} 0, & \text{if } t < 0 \\ t^p, & \text{if } 0 \leq t < t_0 \\ t^{p+1}, & \text{if } t \geq t_0 \end{cases}$$

where $t_0 > 1$ and $p \geq 1$. Denote the class of all such functions F by $\mathcal{F}$. If $F(t) = t$ $\forall t \in \mathbb{R}$, then trivially $F \in \mathcal{F}$.

THEOREM 2.12 ([**223**]). *Let (X, d) be a complete metric space, let $\varphi : X \to \mathbb{R}$ be lower semi-continuous and bounded below, and let $f : X \to X$. Suppose there exists $\eta \in \mathcal{A}$ and $F \in \mathcal{F}$ such that for all $x \in X$,*

$$\eta(d(x, f(x))) \leq F(\varphi(x) - \varphi(f(x))).$$

Then f has a fixed point.

It is shown in Remark 3 of [**106**] that if η is subadditive, then there exists $c > 0$ and $\delta_0 > 0$ such that for any $t \in [0, \delta_0]$, $\eta(t) \geq ct$.

For Theorem 2.12 it is assumed that for $\eta : \mathbb{R}^+ \to \mathbb{R}^+$ there exists $\bar{\varepsilon} > 0$ and $\gamma \in \Gamma$ such that if $\eta(t) \leq \bar{\varepsilon}$, then $\eta(t) \geq \gamma(t)$.

Therefore it appears that if one takes $\eta = \gamma$ and $F(t) = t$ in Theorem 2.12 one obtains Khamsi's result for subadditive η. It is not obvious to us that one fully recovers Khamsi's theorem.

QUESTION. Is it possible to derive the theorem of Zhang and Jiang from the Brézis–Browder order principle?

CHAPTER 3

Nonexpansive Mappings and Zermelo's Theorem

3.1. Introduction

An extension of a theorem attributed variously to Zermelo, Bourbaki, and Kneser provides the basis for Mańka's proof that Caristi's theorem holds in ZF. In the sequel we shall simply refer to this theorem as Zermelo's theorem. This theorem should NOT be confused with the celebrated well-ordering theorem also due to Zermelo, which is equivalent to the Axiom of Choice. See A.3 and A.9 of [**107**] for a brief discussion of constructive aspects of mathematics.

THEOREM 3.1 (Zermelo [**222**]). *Let $(E, \leq)$ be a partially ordered set and let $f : E \to E$ satisfy $x \leq f(x) \; \forall x \in E$. Suppose every chain in $(E, \leq)$ has a least upper bound. Then f has a fixed point in E. In fact, given $x \in E$ it is possible to construct $x^* \in E$ so that $x \leq x^*$ and $f(x^*) = x^*$.*

For a constructive (ZF) proof of this theorem see [**67**, p. 9], [**221**, p. 504], or [**107**, p. 284].

3.2. Convexity Structures

In this section we prove an abstract metric fixed theorem for nonexpansive mappings that contains many known theorems as special cases. Our proof is constructive in that it only relies on Zermelo's theorem. We first need some definitions and we start with a concept inspired by observations of J.-P. Penot in [**169**].

DEFINITION 3.1. A *convexity structure* in a metric space (X, d) is a family Σ of subsets of X such that $\emptyset$, $X \in \Sigma$ and Σ is closed under arbitrary intersections. The structure Σ is said to be [*countably*] *compact* if every [countable] subfamily of Σ which has the finite intersection property has nonempty intersection.

Given a convexity structure Σ in a metric space (X, d), we adopt the following notation: For $D \in \Sigma$ and $x \in X$, set:

$$
\begin{aligned}
r_x(D) &= \sup\{d(x, y) : y \in D\}; \\
r_X(D) &= \inf\{r_x(D) : x \in X\}; \\
r(D) &= \inf\{r_x(D) : x \in D\}.
\end{aligned}
$$

© Springer International Publishing Switzerland 2014
W. Kirk, N. Shahzad, *Fixed Point Theory in Distance Spaces*,
DOI 10.1007/978-3-319-10927-5_3

DEFINITION 3.2. A convexity structure Σ in X is said to be *normal* if given $D \in \Sigma$, $diam(D) > 0 \Rightarrow r(D) < diam(D)$.

A subset A of a metric space X is said to be *admissible* if A is the intersection of closed balls centered at points of X. Thus

$$A = \bigcap_{i \in I} \{B(x_i; r_i) : x_i \in X, \ r_i \geq 0\}.$$

The set of all admissible subsets of X is denoted by $\mathcal{A}(X)$. Of particular interest in metric fixed point theory is the convexity structure $\mathcal{A}(X)$ consisting of all admissible sets in X. Given any bounded set $A \subseteq X$ we set

$$cov(A) := \bigcap \{D : D \in \Sigma \text{ and } D \supseteq A\}.$$

Clearly $cov(A) \in \mathcal{A}(X)$, and thus $A = cov(A) \Leftrightarrow A \in \mathcal{A}(X)$.

Examples of convexity structures

1. Let Σ be the family of all closed and convex subsets of a given closed convex subset of a Banach space X.
2. Let $\Sigma = \mathcal{A}(B)$ where B is the unit ball in a Banach space X.
3. Let $\Sigma = \mathcal{A}(X)$ where X is a bounded metric space.
4. Let $\Sigma = \mathcal{A}(K)$ where K is a closed convex subset of a complete CAT(0) space (see Chap. 9).

Examples of compact convexity structures

5. The same as Example 1, but with K weakly compact.
6. The same as Example 2, but with B the unit ball in a dual Banach space.
7. The same as Example 3, but with X a hyperconvex metric space (see the next chapter).
8. The same as Example 4.

Examples of convexity structures that are compact and normal

9. The same as Example 5, but with K possessing normal structure [110].
10. The same as Example 6, but with B possessing normal structure.
12. The same as Example 7.
13. The same as Example 4.

We now derive the following theorem as an application of Zermelo's theorem (Theorem 3.1). This provides a constructive proof of the original theorem of Kirk [110] for nonexpansive mappings. The original proof is somewhat shorter, but it uses Zorn's Lemma. (Recall that a mapping f of a metric space (X, d) into itself is *nonexpansive* if $d(f(x), f(y)) \leq d(x, y)$ for all $x, y \in X$.) This proof, taken from [115], was inspired by one given by B. Fuchssteiner in [85].

THEOREM 3.2. *If K is a nonempty bounded subset of a metric space (X,d) for which $\Sigma := \mathcal{A}(K)$ is compact and normal, then every nonexpansive mapping $f : K \to K$ has a fixed point.*

PROOF. Since K is bounded, $K \in \Sigma$.

Step 1. Let

$$\mathcal{M} := \{D \in \Sigma : D \neq \emptyset \text{ and } f : D \to D\},$$

and define $f_1 : \mathcal{M} \to \mathcal{M}$ by setting $f_1(D) = cov(f(D))$, $D \in \mathcal{M}$.

Now introduce the order $\preceq$ on $\mathcal{M}$ by setting $D_1 \preceq D_2 \Leftrightarrow D_2 \subseteq D_1$. Then $D \preceq f_1(D) \ \forall \ D \in \mathcal{M}$. Also, by compactness of Σ every chain $\mathcal{C}$ in $(\mathcal{M}, \preceq)$ has a least upper bound, namely $\bigcap \mathcal{C}$. Therefore by Zermelo's theorem, given $D \in \mathcal{M}$ there exists $D^* \in \mathcal{M}$ such that $D \preceq D^*$ and

$$f_1(D^*) = D^*.$$

In particular $cov(f(D^*)) = D^*$.

Step 2. For $D \in \Sigma$, $D \neq \emptyset$, define

$$R(D) = \left\{ r \geq 0 : D \cap \left(\bigcap_{u \in D} B(u; r) \right) \neq \emptyset \right\}.$$

Then $diam(D) \in R(D)$ so $R(D) \neq \emptyset$. Thus $r(D) := \inf \{r \geq 0 : r \in R(D)\}$ is well defined. Now set

$$C(D) = \left\{ z \in D : z \in \bigcap_{u \in D} B(u; r(D)) \right\}.$$

Clearly $C(D) \in \Sigma$.

Assertion. $C(D) \neq \emptyset$.

Proof. If $r > R(D)$, then by the definition of $R(D)$,

$$C_r(D) := \left\{ z \in D : z \in \bigcap_{u \in D} B(u; r) \right\} \neq \emptyset.$$

We will show that $C(D) = \bigcap_{r > r(D)} C_r(D)$ from which the conclusion will follow by compactness of Σ.

Clearly $C(D) \subseteq C_r(D)$ for each $r > r(D)$ because

$$C(D) = D \cap \left(\bigcap_{u \in D} B(u; r(D)) \right) \subseteq D \cap \left(\bigcap_{u \in D} B(u; r) \right) = C_r(D).$$

Thus $C(D) \subseteq \bigcap_{r > r(D)} C_r(D)$. Now suppose there exists

$$z \in \left(\bigcap_{r > r(D)} C_r(D) \right) \setminus C(D).$$

Then there exists $u \in D$ such that $d(z, u) > r(D)$; hence, there exists r' such that $d(z, u) > r' > r(D)$. But $d(z, u) > r'$ implies $z \notin C_{r'}(D)$—a contradiction.

Given $D \in \mathcal{M}$, construct D^* as in Step 1. It is now possible to define $f_2 : \mathcal{M} \to \mathcal{M}$ by setting $f_2(D) = C(D^*)$. As in Step 1, by Zermelo's theorem there exists $D^{**} \in \mathcal{M}$ such that $f_2(D^{**}) = D^{**}$. This implies that $C(D^{**}) = D^{**}$. However since Σ is normal, $diam(D^{**}) > 0 \Rightarrow C(D^{**})$ is a proper subset of D^{**}. Therefore D^{**} must be a singleton consisting of a fixed point of f. □

We now give a proof of Theorem 3.2 which mimics the Zorn Lemma proof in the original paper of Kirk [110]. This more abstract approach, inspired by observations of Penot [169], is self-contained and somewhat quicker.

ALTERNATE PROOF. Since $\mathcal{A}(K)$ is compact, an application of Zorn's Lemma yields the existence of a set $D \in \mathcal{A}(K)$ which is minimal with respect to being nonempty and mapped into itself by f. Also, $cov(f(D)) \subseteq D$ and $f : cov(f(D)) \to cov(f(D))$, so minimality of D implies

$$D = cov(f(D)).$$

Now assume $diam(D) > 0$ and choose r so that $r(D) < r < diam(D)$. It follows that the set

$$C = \{x \in D : D \subseteq B(x; r)\} \neq \varnothing.$$

Also one can quickly check that

$$C = \bigcap_{x \in D} B(x; r) \bigcap D.$$

This proves that $C \in \mathcal{A}(K)$. Now let $z \in C$. Then if $x \in D$

$$d(f(z), f(x)) \leq d(x, z) \leq r.$$

Therefore $f(x) \in B(f(z); r)$ for every $x \in D$; hence, $f(D) \subseteq B(f(z); r)$ from which $cov(f(D)) \subseteq B(f(z); r)$. But $D = cov(f(D))$, so $D \subseteq B(f(z); r)$. This proves that $f(z) \in C$. Hence $f : C \to C$. However if $z, w \in C$, then $d(z, w) \leq r$, so $diam(C) \leq r < diam(D)$. This proves that C is a proper subset of D. Since $C \in \mathcal{A}(K)$ and $f : C \to C$ this contradicts the minimality of D. We thus conclude that $diam(D) = 0$, so D consists of a single point which must be a fixed point of f. □

REMARK 3.1. *In* [114] *it is shown that countable compactness of Σ is sufficient for the validity of Theorem 3.2. However it has since been shown by Kulesza and Lim* [136] *that if (X, d) a bounded metric space for which $\mathcal{A}(X)$ is countable compact and normal, then $\mathcal{A}(X)$ is in fact compact. See* [107, *p. 109] for full details.*

CHAPTER 4

Hyperconvex Metric Spaces

We only briefly discuss this topic because metric fixed point theory in these spaces has been discussed extensively elsewhere (see, e.g., [72] or [107, Chapter 4]). However, since some of the spaces we discuss below are hyper-convex (in particular the so-called $\mathbb{R}$-trees) we touch on a few of the relevant properties of these spaces.

A metric space M is said to be *injective* if it has the following extension property: Whenever Y is a subspace of a metric space X and $f : Y \to M$ is nonexpansive, then f has a nonexpansive extension $\tilde{f} : X \to M$. This fact has several nice consequences. For example, suppose M is injective and suppose M is a subspace of a metric space X. Then since the identity mapping $I : M \to M$ is nonexpansive then I can be extended to a nonexpansive mapping $R : X \to M$. Since R is a retraction of X onto M we have the following.

THEOREM 4.1. *An injective metric space is a nonexpansive retract of any metric space in which it is metrically embedded.*

In light of the above it is clear that an injective metric space must be complete because it is a nonexpansive retract (hence a closed subspace) of its own completion.

DEFINITION 4.1. A metric space (X, d) is said to be *hyperconvex* if for any indexed class of closed balls $B(x_i; r_i)$, $i \in I$, of X which satisfy

$$d(x_i, x_j) \le r_i + r_j \qquad i, j \in I,$$

it is necessarily the case that $\bigcap_{i \in I} B(x_i; r_i) \ne \varnothing$.

It is easy to see that a hyperconvex metric space X is always complete. Also if $x, y \in X$ then there exists $z \in X$ such that $d(x, z) + d(z, y) = d(x, y)$. Thus X is Menger convex. Therefore by Menger's theorem each two points of X are the endpoints of a metric segment.

Of particular relevance is the fact that hyperconvex spaces are injective. Indeed, the following well-known theorem is due to Aronszajn and Panitch-pakdi [12].

© Springer International Publishing Switzerland 2014
W. Kirk, N. Shahzad, *Fixed Point Theory in Distance Spaces*,
DOI 10.1007/978-3-319-10927-5_4

THEOREM 4.2. *A metric space is injective if and only if it is hyperconvex.*

A metric space is said to have the *binary ball intersection property* if any family of closed balls, each two of which intersect, has nonempty intersection. The following is another useful characterization of hyperconvexity. The proof is straightforward.

THEOREM 4.3. *For a complete metric space X the following are equivalent:*

    (1) X is hyperconvex;

    (2) X is metrically convex and has the binary ball intersection property.

For an arbitrary metric space (X, d), J.R. Isbell [99] defined the set of *extremal functions* $\varepsilon(X)$ on X as follows. For any $x \in X$ define $f_x : X \to \mathbb{R}$ by setting

$$f_x(y) = d(x, y), \quad y \in X.$$

The space $\varepsilon(X)$ is the set of all functions $f : X \to \mathbb{R}$ satisfying $f(x) + f(y) \geq d(x, y)$ for all $x, y \in X$, and also satisfying, for some $a \in X$ and all $x \in X$, $f(x) \leq f_a(x)$. The following remarkable fact shows that every metric space can be isometrically embedded in a hyperconvex metric space. (See [99], [72, Section 8], or [107, Section 4.7].)

THEOREM 4.4. *Let (X, d) be a metric space and $\varepsilon(X)$ the set of extremal points on X. Then:*

    1. $\varepsilon(X)$ is a hyperconvex metric space with metric $\rho(f, g) = \sup_{x \in X} |f(x) - g(x)|$ for $f, g \in \varepsilon(X)$.

    2. X is isometrically embedded in $\varepsilon(X)$ via the mapping $f_x : X \to \varepsilon(X)$ defined by $f_x(y) = d(x, y), \quad y \in X$.

    3. If X is isometrically embedded in any hyperconvex metric space Y, then $\varepsilon(X)$ can also be isometrically embedded into Y.

Other useful facts include the following. (Here we use the terminology and notation of the previous chapter.)

PROPOSITION 4.1. *Suppose (X, d) is a bounded hyperconvex metric space. Then each set $D \in \mathcal{A}(X)$ is itself hyperconvex, and the family $\mathcal{A}(X)$ is a compact and normal convexity structure.*

In view of Theorem 3.2 we now have the following:

THEOREM 4.5. *If (X, d) is a bounded hyperconvex metric space, then every nonexpansive mapping $f : X \to X$ has a fixed point.*

The above theorem is actually a special case of the following much more general result.

THEOREM 4.6 ([17]). *Let (X, d) be a bounded hyperconvex metric space, and let $\mathfrak{F}$ be a commuting family of nonexpansive mappings of X into X. Then the common fixed point set of $\mathfrak{F}$ is nonempty and hyperconvex.*

CHAPTER 5

Ultrametric Spaces

5.1. Introduction

Most of the results of this section are taken from Kirk–Shahzad [**127**]. We begin by recalling three definitions of an ultrametric space.

The classical definition goes back over 50 years. See [**138**] for a discussion. A metric space (X, d) is called an *ultrametric space* if the metric d satisfies the strong triangle inequality; namely for all $x, y, z \in X$:

$$d(x, y) \leq \max\{d(x, z), d(y, z)\}.$$

In this case d is said to be *non-archimedean*.

A second definition was inspired by the study of functional analysis in vector valued spaces (cf., [**160**, **175**]). Let X be a nonempty set and let $(\Gamma, \leq)$ be a totally ordered set with $0 \in \Gamma$ and $0 = \min \Gamma$. A mapping $d : X \times X \to \Gamma$ is said to be an *ultrametric distance* on X if for all $x, y, z \in X$

(D_1) $d(x, y) = 0 \Leftrightarrow x = y$;
(D_2) $d(x, y) = d(y, x)$;
(D_3) $d(x, y) \leq \max\{d(x, z), d(z, y)\}$.

A third definition is found in [**33**]. It coincides with the definition given above, except that Γ is assumed to be a complete lattice with least element 0 and a greatest element 1 and (D_3) becomes $d(x, y) \leq \sup\{d(x, z), d(z, y)\}$.

NOTE. In keeping with the metric spirit of this text we choose to use the classical definition, although most of these results hold in more abstract settings.

We first observe that in an ultrametric space *all triangles are isosceles, with the two equal sides at least as long as the third side.* To see this, let x, y, z be elements of an ultrametric space (X, d) with $d(z, y) \geq d(x, z)$, and suppose

$$d(x, y) < \max\{d(x, z), d(z, y)\}.$$

Then $d(x, z) = d(z, y)$, because otherwise

$$d(z, y) > d(x, z) \Rightarrow d(z, y) > \max\{d(x, y), d(x, z)\}.$$

© Springer International Publishing Switzerland 2014
W. Kirk, N. Shahzad, *Fixed Point Theory in Distance Spaces*,
DOI 10.1007/978-3-319-10927-5_5

We use the notation $B(x;r)$ to denote the closed ball

$$B(x;r) = \{y \in X : d(x,y) \leq r\},$$

where $r \geq 0$ (with $B(x;0) = \{x\}$) and we observe that always $diam(B(x;r)) \leq r$.

Another characteristic property of ultrametric spaces is the following:

(5.1) $\alpha \leq \beta$ and $B(x;\alpha) \cap B(y;\beta) \neq \emptyset \Rightarrow B(x;\alpha) \subseteq B(y;\beta).$

Moreover if $\alpha = d(x,y)$, $B(x;\alpha) = B(y;\alpha)$.

DEFINITION 5.1. An ultrametric space (X,d) is said to be *spherically complete* if every chain of balls in X has nonempty intersection.

REMARK 5.1. *An immediate consequence of (5.1) is the fact that $\cap \mathcal{F} \neq \emptyset$ for any family $\mathcal{F}$ of closed balls in a spherically complete ultrametric space which has the property that each two members of $\mathcal{F}$ intersect.*

Ultrametric spaces[1] and hyperconvex metric spaces share many common properties, yet they are quite different in very distinctive ways. The most striking similarity has to do with the injective extension property; the most striking difference is likely the fact that while hyperconvex metric spaces are always metrically convex, ultrametric spaces never are.

An ultrametric space (M,d) is said to have the *extension property* (EP) if given any ultrametric space (X,ρ) and any subspace Y of X, every nonexpansive mapping $f : Y \to M$ has a nonexpansive extension $f' : X \to M$.

The following characterization of spherical completeness is found in [**175**].

THEOREM 5.1. *An ultrametric space is spherically complete if and only if it has the extension property.*

PROOF. ($\Rightarrow$) Suppose (M,d) is spherically complete, let (X,ρ) be an ultrametric space, let Y be a subspace of X, and suppose $f : Y \to M$ is nonexpansive. Let $z \in X$ with $z \notin Y$ and let $Y' = Y \cup \{z\}$. We first show that f has a nonexpansive extension $f' : Y' \to M$.

Now let $\mathcal{F} = \{B(f(y);\rho(y,z)) : y \in Y\}$. We assert that each two members of $\mathcal{F}$ intersect. Indeed, suppose $y_1, y_2 \in Y$ with $\rho(y_1,z) \leq \rho(y_2,z)$. Then $z \in B(y_1;\rho(y_1,z)) \cap B(y_2;\rho(y_2,z))$ so by (5.1)

$$B(y_1;\rho(y_1,z)) \subseteq B(y_2;\rho(y_2,z)).$$

Therefore $d(f(y_1), f(y_2)) \leq \rho(y_1,y_2) \leq \rho(y_2,z)$, so $f(y_1) \in B(f(y_2);\rho(y_2,z))$. Since M is spherically complete, $\cap \mathcal{F} \neq \emptyset$, so let $p \in \cap \mathcal{F}$ and define $f'(z) = p$. Then if $f'(y) = f(y)$ for each $y \in Y$, $d(f'(z), f'(y)) \leq \rho(z,y)$ and f' is

---

[1]Ultrametrics arise naturally in the study of non-archimedean analysis; in particular in the study of normed vector spaces over non-archimedean valuation fields (see [**156**, **160**, **213**]).

a nonexpansive extension of f to Y'. The proof of this implication is now completed by using Zorn's lemma as in the extension theorem of Aronszajn and Panitchpakdi.

($\Leftarrow$) Now assume (M, d) has the extension property but is not spherically complete. Then there exists a decreasing family $\{B(x_i; r_i)\}_{i \in I}$ of closed balls in M for which $\bigcap_{i \in I} B(x_i; r_i) = \varnothing$. Let $M' = M \cup \{p\}$ where $p \notin M$ and define a metric ρ on M' as follows. Set $\rho(p, p) = 0$, $\rho(x, y) = d(x, y)$ if $x, y \in M$; otherwise for $x \in M$ set $\rho(x, p) = \rho(p, x) = d(x, x_j)$ where $x \notin B(x_j; r_j)$. By assumption such $j \in I$ must exist. To see that ρ is well defined, notice that if $x \notin B(x_k; r_k)$ then, since these balls are nested, $d(x_j, x_k) < d(x, x_j)$. Thus $d(x, x_j) = d(x, x_k)$.

By the extension property, the identity mapping on M has an extension $f' : M' \to M$. Also if $x_i \notin B(x_j; r_j)$, it must be the case that $B(x_j; r_j) \subseteq B(x_i; r_i)$. Thus

$$d(f'(p), x_i) = d(f'(p), f'(x_i)) \leq \rho(p, x_i) = d(x_i, x_j) \leq r_i,$$

and $f'(p) \in \bigcap_{i \in I} B(x_i; r_i)$. This contradicts the original assumption and completes the proof. $\square$

5.2. Hyperconvex Ultrametric Spaces

An ultrametric space X in the terminology of [33] is said to be *hyperconvex* if it satisfies the following two conditions (where Γ is assumed to be a complete lattice with least element 0 and a greatest element 1):

(H1) For any family $\{B(x_i; \gamma_i)\}_{i \in I}$ of balls

$$B(x_i; \gamma_i) \cap B(x_j; \gamma_j) \neq \emptyset \,\forall\, i, j \in I \Rightarrow \bigcap_{i \in I} B(x_i; \gamma_i) \neq \emptyset.$$

(H2) For all $x, y \in X$ and $\gamma_1, \gamma_2 \in \Gamma$:

$$d(x, y) \leq \sup\{\gamma_1, \gamma_2\} \Rightarrow \exists\, z \in X \text{ such that } d(x, z) \leq \gamma_1 \text{ and}$$
$$d(z, y) \leq \gamma_2 \text{ (i.e., } B(x; \gamma_1) \cap B(y; \gamma_2) \neq \emptyset).$$

We first observe that in the classical setting the second condition is redundant. Indeed, in any classical ultrametric space,

$$d(x, y) \leq \max\{\gamma_1, \gamma_2\} \Leftrightarrow B(x; \gamma_1) \cap B(y; \gamma_2) \neq \emptyset.$$

Consider balls $B(x; \gamma_1)$ and $B(y; \gamma_2)$ with $d(x, y) \leq \max\{\gamma_1, \gamma_2\}$. There are two cases: If $\gamma_1 \leq \gamma_2$, then $x \in B(x; \gamma_1) \cap B(y; \gamma_2)$. On the other hand if $\gamma_2 \leq \gamma_1$, then $y \in B(x; \gamma_1) \cap B(y; \gamma_2)$. In either case $B(x; \gamma_1) \cap B(y; \gamma_2) \neq \emptyset$. Conversely, suppose $B(x; \gamma_1) \cap B(y; \gamma_2) \neq \emptyset$ and let $z \in B(x; \gamma_1) \cap B(y; \gamma_2)$. Then $d(x, z) \leq \gamma_1$ and $d(y, z) \leq \gamma_2$. By the ultrametric triangle inequality, $d(x, y) \leq \max\{d(x, z), d(y, z)\} \leq \max\{\gamma_1, \gamma_2\}$.

Accordingly, we say that an ultrametric space is a classical *hyperconvex ultrametric* space if it satisfies (H1).

A hyperconvex ultrametric space is never hyperconvex in the metric sense. This is because a hyperconvex metric space is always complete, and each two points are joined by a metric segment. In contrast, as we have seen, each three distinct points of an ultrametric space are the vertices of an isosceles triangle.

Next observe that if $B(x; \gamma) \subseteq B(y; \delta)$, then

$$d(x, y) \leq \delta \leq \sup\{\gamma, \delta\}.$$

Hence any descending collection of balls in a classical hyperconvex ultrametric space has nonempty intersection by (5.1), so we conclude that if an ultrametric space is hyperconvex then it is spherically complete.

Now suppose X is a spherically complete ultrametric space and suppose

$$\{B(x_i; \gamma_i)\}_{i \in I}$$

is a family of balls in X satisfying $d(x_i, x_j) \leq \max\{\gamma_i, \gamma_j\}$. Since the real numbers are linearly ordered, it follows that $\{B(x_i; \gamma_i)\}_{i \in I}$ is a nested chain; hence by spherical completeness $\bigcap_{i \in I} B(x_i; \gamma_i) \neq \emptyset$.

THEOREM 5.2. *A classical ultrametric space is hyperconvex in the sense of (H1) if and only if it is spherically complete.*

Since it is well known that hyperconvex metric spaces are injective, the above fact suggests that spherically complete ultrametric spaces should also be injective. This is indeed the case. Following [175] we say that an ultrametric space (X, d) has the *extension property* (EP) if for every ultrametric space (X, ρ) and any subspace Y of X, any nonexpansive mapping $f : Y \to X$ has a nonexpansive extension $f' : X \to X$. The following is a special case of Theorem 1.3 of [175].

THEOREM 5.3. *Let (X, d) be an ultrametric space. Then the following are equivalent:*

(1) *(X, d) is spherically complete;*
(2) *(X, d) has (EP).*

COROLLARY 5.1. *Let Y be a spherically complete subspace of an ultrametric space X. Then Y is a nonexpansive retract of X.*

PROOF. The identity mapping $I : Y \to Y$ has a nonexpansive extension $r : X \to Y$. $\square$

5.3. Nonexpansive Mappings in Ultrametric Spaces

Suppose (X, d) is an ultrametric space and $f : X \to X$ a mapping. We say that ball $B := B(x; r)$ is *minimal f-invariant* if $f : B \to B$ and $d(u, f(u)) = r$ for all $u \in B$. The following theorem was first proved in [170] using Zorn's Lemma. Here we give a constructive proof that seems to be more illuminating. Specifically, the fact that the conclusion holds in every ball of the form $B(x, d(x, f(x)))$ seems to be a new observation.

THEOREM 5.4 (cf., [170]). *Suppose (X, d) is a spherically complete ultrametric space and suppose $f : X \to X$ is nonexpansive. Then every ball of the form*

$$B(x; d(x, f(x)))$$

contains either a fixed point of f or a minimal f-invariant ball.

PROOF. ([127]) Let $z \in X$, let $r = d(z, f(z))$ and let $u \in B(z; r)$. We assert that $f(u) \in B(z; r)$ and $d(u, f(u)) \leq d(z, f(z))$. To see this we look at two cases. (i) If $d(u, z) < r$, then, since $d(f(z) f(u)) \leq d(z, u)$, by isosceles triangles it must be the case that $d(z, f(u)) = r$, and again by isosceles triangles $d(u, f(u)) = r$. (ii) If $d(z, u) = r$, then by isosceles triangles, either $d(z, f(u)) = r$ and $d(u, f(u)) \leq r$ or $d(z, f(u)) < r$ and $d(u, f(u)) = r$. Thus in either case, $f(u) \in B(z; r)$ and $d(u, f(u)) \leq r$. Therefore every ball in X of the form $B(z; d(z, f(z)))$ is invariant under f.

Now let $x \in X$. We shall show that $B := B(x, d(x, f(x)))$ contains either a fixed point of f or a minimal f-invariant ball. We proceed by induction. Let $x_1 = x$, let $r_1 = d(x_1, f(x_1))$, and set

$$\mu_1 = \inf \{d(x, f(x)) : x \in B(x_1; r_1)\}.$$

Now let $\{\varepsilon_n\}$ be a sequence of positive numbers such that $\lim_{n \to \infty} \varepsilon_n = 0$. If $\mu_1 = r_1$, then $B = B(x_1; r_1)$ is either a singleton or a minimal f-invariant ball, and we are finished. If $r_1 > 0$ and $\mu_1 < r_1$ select $x_2 \in B(x_1; r_1)$ so that

$$r_2 := d(x_2, f(x_2)) < \min \{r_1, \mu_1 + \varepsilon_1\}.$$

Having defined $x_n \in X$, let

$$\mu_n = \inf \{d(x, f(x)) : x \in B(x_n; r_n)\}.$$

As seen above when $n = 1$, if $\mu_n = r_n$ or $r_n = 0$ we are finished. Otherwise select $x_{n+1} \in B(x_n; r_n)$ so that

$$r_{n+1} := d(x_{n+1}, f(x_{n+1})) < \min \{r_n, \mu_n + \varepsilon_n\}.$$

Either this process terminates and the conclusion follows after a finite number of steps, or $\{B(x_n; r_n)\}_{n=1}^{\infty}$ is a nested sequence of nontrivial balls. In the latter case, since X is spherically complete,

$$\bigcap_{n=1}^{\infty} B(x_n; r_n) \neq \emptyset.$$

Since $\{r_n\}$ is decreasing, $r := \lim_{n \to \infty} r_n$ exists. Also $\{\mu_n\}$ is nondecreasing and bounded above, so $\mu := \lim_{n \to \infty} \mu_n$ also exists. Let $x^* \in \cap_{n=1}^{\infty} B(x_n; r_n)$. Then for each n,

$$d(x^*, f(x^*)) \leq \max \{d(x^*, x_n), d(x_n, f(x^*))\} \leq r_n.$$

Moreover, $x^* \in B(x_{n+1}; r_{n+1}) \forall n \Rightarrow$

$$\mu_n \leq d(x^*, f(x^*)) \leq r \leq r_{n+1} \leq \mu_n + \varepsilon_n.$$

Letting $n \to \infty$ we see that $d(x^*, f(x^*)) = \mu = r$. On the other hand,

$$\inf \{d(x, f(x)) : x \in B(x^*; d(x^*, f(x^*)))\} \geq \mu_n,$$

and this implies

$$r \geq \inf \{d(x, f(x)) : x \in B(x^*; d(x^*, f(x^*)))\} \geq \mu = r.$$

If $r > 0$, $B(x^*; d(x^*, f(x^*)))$ is a minimal f-invariant ball contained in B. If $r = 0$, x^* is a fixed point of f.   □

REMARK 5.2. *The above proof requires only that descending **sequences** of closed balls have nonempty intersection.*

REMARK 5.3. *If $B(x; r)$ is a minimal f-invariant ball, then*

$$d\left(f^n(x), f^{n+1}(x)\right) = r$$

for all $n \in \mathbb{N}$.

COROLLARY 5.2 (cf., [**174**]). *Suppose (X, d) is a spherically complete ultrametric space and suppose $f : X \to X$ is strictly contractive $(d(f(x), f(y)) < d(x, y)$ when $x \neq y)$. Then f has a unique fixed point.*

Notice that in Corollary 5.2 the fixed point of f must lie in *every* ball of the form $B(x; d(x, f(x)))$ for $x \in X$. Hence these balls are nested, and consequently

$$\{x^*\} = \bigcap_{x \in X} B(x; d(x, f(x))),$$

where $f(x^*) = x^*$. Also, if $x \in X$ and $x \neq x^*$, then $d(x^*, f(x)) < d(x^*, x) \Rightarrow d(x^*, x) = d(x, f(x))$. This suggests a method for approximating the fixed point of a strictly contractive mapping.

5.4. Structure of the "Fixed Point Set" of Nonexpansive Mappings

In this section we examine the nature of the "fixed point set" under the assumptions of Theorem 5.4.

THEOREM 5.5. *Suppose (X, d) is a spherically complete ultrametric space and suppose $f : X \to X$ is nonexpansive. Let*

$$M = \{x \in X : \exists \, r \geq 0 \text{ such that } d(u, f(u)) = r \, \forall u \in B(x; r)\}.$$

Then M is spherically complete, and hence a nonexpansive retract of X.

PROOF. Let $B(x_i; \gamma_i)$ be a descending collection of closed balls centered at points $x_i \in M$. Then for each i there exists $r_i \geq 0$ such that $d(u, x_i) \leq r_i \Rightarrow d(u, f(u)) = r_i$. Since X is spherically complete, $B := \bigcap_{i \in I} B(x_i; \gamma_i) \neq \emptyset$. If $\gamma_i \leq r_i$ for some i, then the collection of balls all eventually lie in $B(x_i; r_i)$, and so $B \subseteq B(x_i; r_i) \subseteq M$. So, suppose $r_i < \gamma_i$ for each i. Let $x \in B$. Then $d(f(x), f(x_i)) \leq d(x, x_i) \leq \gamma_i$. Also,

$$d(f(x), x_i) \leq \max \{d(f(x), f(x_i)), d(f(x_i), x_i)\} \leq \max \{\gamma_i, r_i\} = \gamma_i.$$

Thus $f : B \to B$. But B is itself spherically complete. So $B \cap M \neq \emptyset$. This proves that M is spherically complete. The fact that M is a nonexpansive retract of X follows from Corollary 5.1. $\square$

With f and M as in Theorem 5.5, suppose $x \in M$, and suppose there exists $r > 0$ such that $f : B(x; r) \to B(x; r)$ and $d(u, f(u)) = r \ \forall \ u \in B(x; r)$. Then $d(u, x) < r \Rightarrow d(f(u), x) = r$. Moreover, since $d(f^n(x), f^{n+1}(x)) = r$ for all $n \in \mathbb{N}$, by isosceles triangles we have $d(x, f^n(x)) < r \Rightarrow d(x, f^{n+1}(x)) = r$ for any $n \in \mathbb{N}$. The simple example below shows that this behavior is typical.

Example. Let $X = \{a, b, c, d\}$ with $d(x, x) = 0$ for all $x \in X$; $d(a, b) = d(c, d) = 1/2$; $d(a, c) = d(a, d) = d(b, c) = d(b, d) = 1$; $d(y, x) = d(x, y)$ for all $x, y \in X$. Then (X, d) is a spherically complete ultrametric space. Define $f(a) = c$; $f(c) = a$; $f(b) = d$; $f(d) = b$. Then f is nonexpansive, f does not have any fixed points, and $M = X$.

REMARK 5.4. *Under the assumptions of Theorem 5.5, if $z \in X$ satisfies $d(z, f(z)) = \inf \{d(x, f(x)) : x \in X\}$, then*

$$d(u, f(u)) = d(z, f(z))$$

for all $u \in B(z; d(z, f(z)))$.

Suppose $x, y \in M$ with $d(x, f(x)) = r_1$ and $d(y, f(y)) = r_2$. If $B(x; r_1) \cap B(y; r_2) = \emptyset$, then $d(x, y) := d > \max \{r_1, r_2\}$. By isosceles triangles,

$$d(f(x), f(y)) = d.$$

On the other hand if $B(x; r_1) \cap B(y; r_2) \neq \emptyset$, then $r_1 = r_2$. Thus if $x, y \in M$, either $d(x, f(x)) = d(y, f(y)) = r$ and $d(x, y) \leq r$ or

$$d(x, y) > \max \{d(x, f(x)), d(y, f(y))\}$$

and $d(x, y) = d(f(x), f(y))$.

5.5. A Strong Fixed Point Theorem

The existence part of the following theorem is Theorem 1 of [**179**]. The proof given there is indirect and also relies on Zorn's Lemma. As in Theorem 5.4, we give a more constructive proof here, one that also shows $f : X \to X$ has a fixed point in *every* ball of the form $B(x; d(x, f(x)))$. The assumption that f is *strictly contracting on orbits* [**178**] means that $f(x) \neq x$ implies $d\left(f^2(x), f(x)\right) < d\left(f(x), x\right)$ for each $x \in X$.

THEOREM 5.6. *Let (X, d) be a spherically complete ultrametric space, let $f : X \to X$, and assume the following properties are satisfied:*

(1) If $z \neq f(z)$ and $d(x, f(z)) \leq d\left(f^2(z), f(z)\right)$, then

$$d(x, f(x)) \leq d(z, f(z)).$$

(2) f is strictly contracting on orbits.

Then f has a fixed point in every ball of the form $B(x; d(x, f(x)))$.

PROOF. Choose $x \in X$. We shall show that $B(x; d(x, f(x)))$ contains a fixed point of f. Let Ω denote the set of all countable ordinals and let $x_1 = x$. We proceed by transfinite induction. Let $\beta \in \Omega$ and assume x_α has been defined for all $\alpha < \beta$, where $\{B(x_\alpha; d(x_\alpha, f(x_\alpha)))\}_{\alpha < \beta}$ is a descending chain of balls and $\{d(x_\alpha, f(x_\alpha))\}_{\alpha < \beta}$ a descending chain of real numbers. If $x_{\alpha'} = f(x_{\alpha'})$ for some $\alpha' < \beta$, take $x_{\alpha'} = x_\beta$. Otherwise, if $\beta = \alpha + 1$, take $x_\beta = f(x_\alpha)$. If β is a limit ordinal, choose

$$x_\beta \in \bigcap_{\alpha < \beta} B(x_\alpha; d(x_\alpha, f(x_\alpha))).$$

We may now assume that $x_\alpha \neq f(x_\alpha)$ for all $\alpha < \beta$; otherwise x_β is a fixed point of f in $B(x; d(x; f(x)))$ and there is nothing more to prove. Suppose $\beta = \alpha + 1$. Then by (2) we have $d(x_\beta, f(x_\beta)) = d\left(f(x_\alpha), f^2(x_\alpha)\right) < d(x_\alpha, f(x_\alpha))$. Since

$$f(x_\alpha) \in B(x_\beta; d(x_\beta, f(x_\beta))) \cap B(x_\alpha; d(x_\alpha, f(x_\alpha))),$$

it must be the case that

$$B(x_\beta; d(x_\beta, f(x_\beta))) = B(x_{\alpha+1}; d(x_{\alpha+1}, f(x_{\alpha+1}))) \subset B(x_\alpha; d(x_\alpha, f(x_\alpha))).$$

If β is a limit ordinal, then $x_\beta \in \bigcap_{\alpha < \beta} B(x_\alpha; d(x_\alpha, f(x_\alpha)))$, and in particular

$$x_\beta \in B(x_{\alpha+1}; d(x_{\alpha+1}, f(x_{\alpha+1}))) = B\left(f(x_\alpha); d\left(f(x_\alpha), f^2(x_\alpha)\right)\right).$$

Thus $d(x_\beta, f(x_\alpha)) \leq d\left(f(x_\alpha), f^2(x_\alpha)\right)$. Since we are assuming $x_\alpha \neq f(x_\alpha)$, condition (1) implies

$$d(x_\beta, f(x_\beta)) \leq d(x_\alpha, f(x_\alpha))$$

for each $\alpha < \beta$. Also

$$x_\beta \in B(x_\beta; d(x_\beta, f(x_\beta))) \cap \bigcap_{\alpha < \beta} B(x_\alpha; d(x_\alpha, f(x_\alpha)))$$

so it must be the case that $x_\beta \in B(x_\beta; d(x_\beta, f(x_\beta))) \subseteq B(x_\alpha; d(x_\alpha, f(x_\alpha)))$ for each $\alpha < \beta$; hence

$$B(x_\beta; d(x_\beta, f(x_\beta))) \subseteq \bigcap_{\alpha < \beta} B(x_\alpha; d(x_\alpha, f(x_\alpha))).$$

We have thus defined x_α for all $\alpha \in \Omega$. Moreover the transfinite sequence

$$\{d(x_\alpha, f(x_\alpha))\}_{\alpha \in \Omega}$$

is nonincreasing. Also, by (2), $d(x_\alpha, f(x_\alpha)) > 0$ implies

$$0 \leq d(x_{\alpha+1}, f(x_{\alpha+1})) = d(f(x_\alpha), f^2(x_\alpha)) < d(x_\alpha, f(x_\alpha)).$$

Observe that $x_\alpha \neq f(x_\alpha)$ is not possible for all $\alpha \in \Omega$ because otherwise it follows from $\alpha' < \alpha$ that $d(x_{\alpha'}, f(x_{\alpha'})) > d(x_\alpha, f(x_\alpha))$. Then, since Ω has cofinal type ω_1, the transfinite sequence $\{d(x_\alpha, f(x_\alpha))\}_{\alpha \in \Omega}$ of real positive numbers ($\neq 0$) would be of coinitial type ω_1, whereas the coinitial type of $\{r \in \mathbb{R} : r > 0\}$ is countable. $\qquad \square$

We now have the following extension of Corollary 5.2.

COROLLARY 5.3. *Let (X, d) be a spherically complete ultrametric space. Suppose $f : X \to X$ is nonexpansive and strictly contracting on orbits. Then f has a fixed point.*

PROOF. If f is nonexpansive on X, if $z \neq f(z)$ for $z \in X$, and if $d(x, f(z)) \leq d(f^2(z), f(z))$, then, since $d(f^2(z), f(z)) \leq d(f(z), z)$, Condition (1) of Theorem 5.6 holds. $\qquad \square$

The above corollary shows that every strictly contractive mapping defined on a spherically complete ultrametric space has a unique fixed point. In fact, the following is true.

PROPOSITION 5.1. *Let (X, d) be an ultrametric space. Then the following are equivalent:*

(a) *X is spherically complete.*
(b) *Every strictly contractive mapping $f : X \to X$ has a fixed point.*

PROOF. (a) $\Rightarrow$ (b): This is a special case of Corollary 5.3.

(b) $\Rightarrow$ (a): (cf., Lemma 2 (b) in [180]). Assume that X is not spherically complete. Then there is a strictly decreasing family $\{B(a_\iota; \gamma_\iota)\}_{\iota < \lambda}$ of balls such that $\bigcap_{\iota < \lambda} (B(a_\iota; \gamma_\iota)) = \emptyset$. We may further assume λ is a limit ordinal and that $\{\gamma_\iota\}_{\iota < \lambda}$ is strictly decreasing. Set $B_\iota = B(a_\iota; \gamma_\iota)$ for $\iota < \lambda$. For each $x \in X$ there is a smallest ordinal $\kappa(x) < \lambda$ such that $x \notin B_{\kappa(x)}$. Define $f : X \to X$ by setting $f(x) = a_{\kappa(x)}$.

To see that f is strictly contractive, let $x, y \in X$ with $x \neq y$. If $\kappa(x) = \kappa(y)$, then

$$d(f(x), f(y)) = d(a_{\kappa(x)}, a_{\kappa(y)}) = 0 < d(x, y).$$

Now suppose $\kappa(x) < \kappa(y)$. Then $B_{\kappa(y)} \subset B_{\kappa(x)}$; thus, $x \notin B_{\kappa(x)}$ and $y \in B_{\kappa(x)}$. Hence

$$d(x, y) > \gamma_{\kappa(x)} \geq d(a_{\kappa(x)}, a_{\kappa(y)}) = d(f(x), f(y)).$$

Thus f is strictly contractive. Clearly f has no fixed point. $\qquad \square$

REMARK 5.5. *Implicit in the proof of Theorem 5.6 is a transfinite method for "approximating" the fixed points of f. Let (X, d) be spherically complete and $f : X \to X$ strictly contractive. Then f has a unique fixed point z in X. Now let $x \in X$ and construct the transfinite sequence $\{x_\alpha\}$ as follows. Let Ω denote the set of all countable ordinals and let $x_1 = x$. We proceed by transfinite induction. Let $\beta \in \Omega$ and assume x_α has been defined for all $\alpha < \beta$, where $\{B(x_\alpha; d(x_\alpha, f(x_\alpha)))\}_{\alpha < \beta}$ is a descending chain of balls and $\{d(x_\alpha, f(x_\alpha))\}_{\alpha < \beta}$ a descending chain of real numbers. If $x_{\alpha'} = f(x_{\alpha'})$ for some $\alpha' < \beta$ take $x_{\alpha'} = x_\beta$. Otherwise, if β is not a limit ordinal, say $\beta = \alpha + 1$, take $x_\beta = f(x_\alpha)$. If β is a limit ordinal and*

$$\bigcap_{\alpha < \beta} B(x_\alpha; d(x_\alpha, f(x_\alpha))) = \{z\},$$

define $x_\beta = z$. Otherwise choose

$$x_\beta \in \bigcap_{\alpha < \beta} B(x_\alpha; d(x_\alpha, f(x_\alpha))) \setminus \{z\}.$$

This sequence must eventually be constant. Let μ be the smallest ordinal such that $x_{\mu+1} = x_\mu$. If μ is not a limit ordinal, then the transfinite sequence terminates at x_μ, and $x_\mu = f(x_\mu) = z$. If μ is a limit ordinal, then $\bigcap_{\gamma < \mu} B(x_\gamma; d(x_\gamma, f(x_\gamma))) = \{z\}$. See [181] *for more details.*

We now state two facts which are special cases of more abstract results proved elsewhere. An ultrametric space (X, d) is said to be an *immediate extension* of an ultrametric space (Y, d) if $Y \subseteq X$ and if for each $x \in X$ and every $y \in Y$ with $y \neq x$ there exists $y' \in Y$ such that $d(y', x) < d(y, x)$.

THEOREM 5.7 ([177]). *Every ultrametric space (X, d) has an immediate extension which is spherically complete. (This space is called the spherical completion of X.)*

THEOREM 5.8 ([176]). *Let Y be a subspace of a spherically complete metric space (X, d), and suppose $f : Y \to Y$ is strictly contractive. Then there exists $f' : X \to X$ such that f' is strictly contractive and extends f.*

REMARK 5.6. *Most of the results described in this chapter hold in more abstract settings where the distance function d takes values in an ordered set Γ. In some cases, especially when Γ is totally ordered, the arguments parallel the ones given here and in other cases more technical arguments are needed. Although we emphasize the metric approach here, it is only appropriate to mention the more general approach. We follow the terminology of* [180].

DEFINITION 5.2. *Let $(\Gamma, \leq)$ be an ordered set with smallest element 0, and let X be a nonempty set. A mapping $d : X \times X \to \Gamma$ is called an ultrametric distance function if for all $x, y, z \in X$*

$(d\ 1)\quad d(x,y) = 0 \Leftrightarrow x = y;$

$(d\ 2)\quad d(x,y) = d(y,x);$

$(d\ 3)\quad d(x,y) \leq \gamma$ and $d(y,z) \leq \gamma \Rightarrow d(x,z) \leq \gamma$ for all $\gamma \in \Gamma.$

In this setting (X, d, Γ) is called an ultrametric space and $d(x,y)$ is called the ultrametric distance between x and y. If Γ is totally ordered, then $(d\ 3)$ becomes

$(d\ 3')\quad d(x,z) \leq \max\{d(x,y), d(y,z)\}.$

Several examples are discussed in [**180**], some where Γ is totally ordered and others where Γ is not totally ordered.

5.6. Best Approximation

A subspace A of a metric space is said to be an *almost nonexpansive retract* of X if for any $\lambda > 1$ there exists a retraction r_λ of X onto A such that r_λ is λ-Lipschitz, i.e., $d(r_\lambda(x), r_\lambda(z)) \leq \lambda d(x,z)$ for all $x, z \in X$.

THEOREM 5.9. *A metric space X is ultrametric if and only if every closed subset A of X is an almost nonexpansive retract of X.*

One implication of this result is Theorem 2.9 of [**37**]. The other implication is alluded to in the lecture notes [N. Brodskiy, Asymptotic Dimension of Groups], which are based on [**37**]. An analysis of the proof of Theorem 2.9 of [**37**] leads to the following.

THEOREM 5.10. *Suppose K is a spherically complete subspace of an ultrametric space X, and suppose $f : K \to X$ satisfies $d(f(x), f(y)) \leq k d(x,y)$ for each $x, y \in K$, where $k \in (0,1)$. Then for any $\mu > 1$ there exists $x^* \in K$ such that*

$$d(f(x^*), x^*) \leq \mu \, dist(f(x^*), K).$$

PROOF. Given $k \in (0,1)$, choose $\lambda > 1$ so that $k\lambda < 1$, and choose $\delta > 1$ so that $\delta \leq \mu$ and $\delta^2 < \lambda$. As seen in the proof of Theorem 2.9 of [**37**] it is possible to define an order $\prec$ on X such that for every nonempty bounded subset C of X the restricted order $\prec|_C$ is a well-order. Define a retraction $r : X \to K$ as follows. For $x \in X$ let $B_x = \{b \in K : d(x,b) \leq \delta \, dist(x,K)\}$ and take $r(x)$ to be the point of B_x which is minimal with respect to $\prec$. It is shown in the proof of Theorem 2.9 of [**37**] that r is λ-Lipschitz. For $x \in K$, set $g(x) = r \circ f(x)$. Then $g : K \to K$, and moreover for each $x, y \in K$,

$$
\begin{aligned}
d(g(x), g(y)) &= d(r \circ f(x), r \circ f(y)) \\
&\leq \lambda d(f(x), f(y)) \\
&\leq k\lambda d(x,y),
\end{aligned}
$$

so by Corollary 5.2 g has a unique fixed point x^* in K. Since $x^* = r \circ f(x^*) \in B_{f(x^*)}$ we have

$$d(f(x^*), x^*) = d(f(x^*), r \circ f(x^*)) \leq \delta \, dist(f(x^*), K) \leq \mu \, dist(f(x^*), K).$$

$\square$

The following is a consequence of results of [**147**]. For the sake of completeness we give the simple proof. Recall that a subspace Y of a metric space X is *proximinal* in X if for any $z \in X$ there exists $y \in Y$ such that $d(z, y) = dist(z, Y)$.

THEOREM 5.11. *Let Y be a spherically complete subspace of an ultrametric space (X, d). Then Y is proximinal in X.*

PROOF. Let $z \in X \backslash Y$, and let $d = dist(z, Y)$. Choose $x_n \in Y$ so that $d_n := d(x_n, z) \leq d + \frac{1}{n}$ and so that $\{d_n\}$ is nonincreasing. Then $m > n \Rightarrow$ $d(x_n, x_m) = d_n$. Hence $\{x_n, x_{n+1}, \cdots\} \subset B(x_n; d_n) \cap Y$. So $\{B(x_n; d_n)\}$ is a descending sequence of nonempty balls in Y. Since Y is spherically complete, there exists $x \in \cap_{n=1}^{\infty} B(x_n; d_n) \cap Y$. Clearly $d(x, z) = d$. □

THEOREM 5.12. *Let Y be a spherically complete subspace of an ultrametric space X, and let $x^* \in X \backslash Y$. Suppose $f : X \to X$ is a mapping for which $f(x^*) = x^*$. Also assume that f is nonexpansive on $Y \cup \{x^*\}$ and that Y is f-invariant. Then f has a fixed point in Y which is a nearest point of x^* in Y, or Y contains a minimal f-invariant set, each point of which is a nearest point to x^* in Y.*

PROOF. Let $d = dist(x^*, Y)$ and let $Z = B(x^*; d) \cap Y$. By Theorem 5.11, Z is nonempty. Let $y \in Z$. Then

$$
\begin{aligned}
dist(x^*, Y) &\leq d(x^*, f(y)) \\
&\leq d(x^*, y) \\
&= dist(x^*, Y).
\end{aligned}
$$

This implies that Z is f-invariant. Now let $y \in Z$. Since

$$d(y, f(y)) \leq \max\{d(y, x^*), d(f(y), x^*)\} = d,$$

it must be the case that

$$B(y; d(y, f(y))) \cap Y \subseteq B(x^*; d).$$

Therefore $B(y; d(y, f(y))) \cap Y \subseteq Z$, and

$$f(B(y; d(y, f(y))) \cap Y) \subseteq B(y; d(y, f(y))) \cap Y.$$

By Theorem 5.4, $B(y; d(y, f(y))) \cap Y$ contains either a fixed point y^* of f which is in Z, or a minimal f-invariant ball which necessarily lies in Z. □

COROLLARY 5.4. *Let Y be a spherically complete subspace of an ultrametric space X and suppose $f : X \to X$ is a mapping having a fixed point $x^* \in X \backslash Y$. Assume that f is strictly contractive on $Y \cup \{x^*\}$ and Y is f-invariant. Then there exists a unique fixed point y^* of f which is a nearest point of x^* in Y.*

PROOF. Let $d = dist(x^*, Y)$ and let $Z = B(x^*; d) \cap Y$. As we have seen, $Z \neq \emptyset$ and $f : Z \to Z$. If $y^* \in Z$ and $f(y^*) \neq y^*$, then we have the contradiction $dist(x^*, Y) \leq d(x^*, f(y^*)) < d(x^*, y^*) = dist(x^*, Y)$. □

Part II

Length Spaces and Geodesic Spaces

CHAPTER 6

Busemann Spaces and Hyperbolic Spaces

We begin with the fundamental definitions. These are taken from [**166**].

DEFINITION 6.1. Let X be a metric space. A *geodesic path* (or simply a *geodesic*) in X is a path $\gamma : [a, b] \to X$, where γ is an isometry. A *geodesic ray* is an isometry $\gamma : \mathbb{R}^+ \to X$, and a *geodesic line* is an isometry $\gamma : \mathbb{R} \to X$.

DEFINITION 6.2. Let E be a vector space. A subset $X \subset E$ is said to be *affinely convex* if for all $x, y \in X$ the affine segment $[x, y] := \{(1-t)x + ty : t \in [0, 1]\}$ is contained in X.

DEFINITION 6.3. Let E be a vector space and let C be an affinely convex subset of E. Then a function $f : C \to \mathbb{R}$ is said to be *convex* if for every $x, y \in C$ and every $t \in [0, 1]$,

$$f((1-t)x + ty) \le (1-t)f(x) + tf(y).$$

DEFINITION 6.4. A metric space X is said to be a *geodesic space* if given two arbitrary points of X there exists a geodesic path that joins them.

A *Busemann space* (also known as a *Busemann convex space*) is a geodesic metric space X such that for any two geodesics $\gamma : [a, b] \to X$ and $\gamma' : [a', b'] \to X$, the map $D_{\gamma, \gamma'} : [a, b] \times [a', b'] \to \mathbb{R}$ defined by

$$D_{\gamma, \gamma'}(t, t') = d(\gamma(t), \gamma'(t'))$$

is convex. Equivalently, let $[x_0, x_1]$ and $[x'_0, x'_1]$ be two geodesic segments in X. For every $t \in [0, 1]$ let x_t be the point on $[x_0, x_1]$ satisfying $d(x_0, x_t) = td(x_0, x_1)$ and let x'_t be the point on $[x'_0, x'_1]$ satisfying $d(x'_0, x'_t) = td(x'_0, x'_1)$. Then

$$d(x_t, x'_t) \le (1-t)d(x_0, x'_0) + td(x_1, x'_1).$$

The following two conditions are necessary and sufficient conditions for a geodesic metric space X to be a Busemann space.

(1) Let $[x_0, x_1]$ and $[x_0, x'_1]$ be two geodesic segments in X having a common initial point x_0, and let m and m' be their respective midpoints. Then

$$d(m, m') \le \frac{1}{2}[d(x_0, x_1) + d(x_0, x'_1)].$$

© Springer International Publishing Switzerland 2014
W. Kirk, N. Shahzad, *Fixed Point Theory in Distance Spaces*,
DOI 10.1007/978-3-319-10927-5_6

(2) Let $[x_0, x_1]$ and $[x_0', x_1']$ be two geodesic segments in X, and let m and m' be their respective midpoints. Then

$$d(m, m') \leq \frac{1}{2} \left[d(x_0, x_1) + d(x_0', x_1') \right].$$

In a Busemann space the geodesic joining any two points is unique. To see this, let $[x_0, x_1]$ and $[x_0', x_1']$ be two geodesic segments in X. For every $t \in [0, 1]$ let x_t be the point on $[x_0, x_1]$ satisfying $d(x_0, x_t) = td(x_0, x_1)$ and let x_t' be the point on $[x_0', x_1']$ satisfying $d(x_0', x_t') = td(x_0', x_1')$. Then

$$d(x_t, x_t') \leq (1 - t) d(x_0, x_0') + td(x_1, x_1').$$

From this we see that if $x_0 = x_0'$ and $x_1 = x_1'$, then it follows that $x_t = x_t'$ for all $t \in [0, 1]$.

DEFINITION 6.5 ([**133**]). (X, d, W) is called a *hyperbolic space* if (X, d) is a metric space and $W : X \times X \times [0, 1] \to X$ is a function satisfying

(i) $\forall x, y, z \in X$ and $\forall \lambda \in [0, 1]$, $d(z, W(x, y, \lambda)) \leq (1 - \lambda) d(z, x) + \lambda d(z, y)$;

(ii) $\forall x, y \in X$ and $\forall \lambda_1, \lambda_2 \in [0, 1]$, $d(W(x, y, \lambda_1), W(x, y, \lambda_2)) = |\lambda_1 - \lambda_2| d(x, y)$;

(iii) $\forall x, y \in X$ and $\forall \lambda \in [0, 1]$, $W(x, y, \lambda) = W(y, x, 1 - \lambda)$;

(iv) $\forall x, y, z, w \in X$ and $\forall \lambda \in [0, 1]$, $d(W(x, z, \lambda), W(y, w, \lambda)) \leq (1 - \lambda) d(x, y) + \lambda d(z, w)$.

If only condition (i) is satisfied, then (X, d, W) is a convex metric space in the sense of *Takahashi* (cf., [**208**]). We shall use (X, d) for (X, d, W) when there is no ambiguity. All four conditions imply that the space is a Busemann space. Conditions (i)–(iii) are equivalent to (X, d, W) being a space of hyperbolic type in the sense of [**86**]. For these spaces we have the following very useful fact.

THEOREM 6.1. *Let (X, d) be a metric space of hyperbolic type and let K be a bounded subset of X. Suppose $f : K \to X$ is nonexpansive. Fix $\alpha \in (0, 1)$ and define $g : K \to X$ by taking $g(x)$ to be the point of $[x, f(x)]$ satisfying*

$$d(x, g(x)) = \alpha d(x, f(x)), \qquad x \in K.$$

Then if $\{g^n(x)\} \subset K$ for $x \in K$, g is asymptotically regular at x. In particular, if $f : K \to K$, then $\inf \{d(x, f(x)) : x \in K\} = 0$.

In what immediately follows we only use condition (i). We shall adopt the customary notation and write $W(x, y, \lambda) = (1 - \lambda)x \oplus \lambda y$, and we shall say a subset K of X is convex if $x, y \in K \Rightarrow (1 - \lambda)x \oplus \lambda y \in K$ for all $\lambda \in [0, 1]$. Recall that a mapping f from a topological space X into a metric space M is said to be *r-continuous* for $r > 0$ if each point $x \in X$ has a neighborhood U_x such that $diam(f(U_x)) \leq r$.

THEOREM 6.2 ([**122**]). *Let (X, d) be a compact Busemann space (or, more generally, a Takahashi convex space) and suppose $f : X \to X$*

is r-continuous. Then there exists a continuous mapping $\bar{f} : X \to X$ *such that* $d\left(f\left(x\right), \bar{f}\left(x\right)\right) \leq r$ *for each* $x \in X$. *In particular, if* X *has the fixed point property for continuous mappings, there exists* $x_0 \in X$ *such that* $d\left(x_0, f\left(x_0\right)\right) \leq r$.

For $x_1, x_2 \in X$ and $a_1, a_2 \in [0, 1]$ satisfying $a_1 + a_2 = 1$, let $a_1 x_1 \oplus a_2 x_2$ denote the unique point of X for which

$$d\left(x_1, a_1 x_1 \oplus a_2 x_2\right) = a_2 d\left(x_1, x_2\right) \text{ and } d\left(x_2, a_1 x_1 \oplus a_2 x_2\right) = a_1 d\left(x_1, x_2\right).$$

Now for $a_1, a_2, a_3 \in [0, 1]$ with $a_1 + a_2 + a_3 = 1$, and an ordered triple $(x_1, x_2, x_3) \in \prod_{i=1}^{3} X$, define $a_1 x_1 \oplus a_2 x_2 \oplus a_3 x_3 = x_3$ if $a_3 = 1$. Otherwise set

$$a_1 x_1 \oplus a_2 x_2 \oplus a_3 x_3 = a_3 x_3 \oplus (1 - a_3) \left[\frac{a_2}{1 - a_3} x_2 \oplus \frac{a_1}{1 - a_3} x_1 \right].$$

Since the metric d is convex, for each $x \in X$,

$$d\left(x, a_3 x_3 \oplus (1 - a_3) \left[\frac{a_2}{1 - a_3} x_2 \oplus \frac{a_1}{1 - a_3} x_1 \right] \right)$$
$$\leq a_3 d\left(x, x_3\right) + (1 - a_3) d\left(x, \frac{a_2}{1 - a_3} x_2 \oplus \frac{a_1}{1 - a_3} x_1 \right)$$
$$\leq a_3 d\left(x, x_3\right) + a_2 d\left(x, x_2\right) + a_1 d\left(x, x_1\right).$$

Having defined $a_1 x_1 \oplus a_2 x_2 \oplus a_3 x_3 \oplus \cdots \oplus a_{n-1} x_{n-1}$ for $(x_1, \cdots, x_{n-1}) \in \prod_{i=1}^{n-1} X$, $\{a_i\}_{i=1}^{n-1} \subset [0, 1]$, and $\sum_{i=1}^{n-1} a_i = 1$, suppose $(x_1, \cdots, x_n) \in \prod_{i=1}^{n} X$, $\{a_i\}_{i=1}^{n} \subset [0, 1]$, and $\sum_{i=1}^{n} a_i = 1$, and set

$$a_1 x_1 \oplus a_2 x_2 \oplus a_3 x_3 \oplus \cdots \oplus a_n x_n = x_n \text{ if } a_n = 1.$$

Otherwise set

$$a_1 x_1 \oplus a_2 x_2 \oplus a_3 x_3 \oplus \cdots \oplus a_n x_n$$
$$= a_n x_n \oplus (1 - a_n) \left[\frac{a_1}{1 - a_n} x_1 \oplus \frac{a_2}{1 - a_n} x_2 \oplus \cdots \oplus \frac{a_{n-1}}{1 - a_n} x_{n-1} \right].$$

We now adopt the notation

$$a_1 x_1 \oplus a_2 x_2 \oplus a_3 x_3 \oplus \cdots \oplus a_n x_n = \sum_{i=1}^{n} \overrightarrow{a_i x_i}.$$

Observe that with this convention we have for all $x \in X$,

$$d\left(x, \sum_{i=1}^{n} \overrightarrow{a_i x_i} \right) \leq \sum_{i=1}^{n} a_i d\left(x, x_i\right).$$

PROOF OF THEOREM 6.2. Since f is r-continuous, for each $x \in X$ there exists $r_x > 0$ such that $diam\left(f\left(U\left(x; r_x\right)\right)\right) \leq \varepsilon$, where $U\left(x; r_x\right)$ denotes the *open* ball centered at x with radius r_x. Since X is compact there exists a finite set $\{x_1, \cdots, x_j\} \subseteq X$ such that $X \subseteq \cup_{i=1}^{j} U\left(x_i; r_{x_i}/2\right)$. Let $r = \inf\left(r_{x_i} : 1 \leq i \leq j\right)$. We now have the following: If $x, y \in X$ and

$d(x,y) \leq r/2$, then there exists $1 \leq i \leq j$ such that $x, y \in U(x_i; r_{x_i})$. Hence $d(f(x), f(y)) \leq r$.

Again using the fact that X is compact, there exists $A = \{a_1, \cdots, a_n\} \subseteq X$ such that $X \subseteq \cup_{i=1}^{n} U_i$ where $U_i = U(a_i; r/2)$, $i = 1, \cdots, n$. Then $\{U_i\}$ is a finite open covering of X so there exists a partition of unity $\{\phi_i\}_{i=1}^{n}$ of X dominated by the family $\{U_i\}$.

Now define the function $\bar{f} : X \to X$ as follows: $\bar{f}(x) = \sum_{i=1}^{n} \overrightarrow{\phi_i(x) f(a_i)}$ for each $x \in X$. Since each of the functions ϕ_i is continuous, $\bar{f}$ is continuous. Then for $x \in X$,

$$d(f(x), \bar{f}(x)) = d\left(f(x), \sum_{i=1}^{n} \overrightarrow{\phi_i(x) f(a_i)}\right)$$

$$\leq \sum_{i=1}^{n} \phi_i(x) d(f(x), f(a_i)).$$

Since $\phi_i(x) = 0$ if $x \notin U_i$ while $d(f(x), f(a_i)) \leq r$ if $x \in U_i$, we have $d(f(x), \bar{f}(x)) \leq r$. $\square$

6.1. Convex Combinations in a Busemann Space

We now summarize the results of [9]. Throughout this section X denotes a complete Busemann space. We take as our point of departure the approach of the previous section, but with the goal of defining the convex combination of a finite set of points of X that is independent of the order in which they are chosen. This procedure suggests two new ways to define the convex hull of a subset of X. We discuss this in more detail at the end of the section. Our motivation is to try to find a more analytic approach to the study of convex hulls of subsets of Busemann spaces. At this point it appears that we have been only partially successful.

We proceed by induction. Having defined $a_1 x_1 \oplus a_2 x_2$ for $\{x_1, x_2\} \subset X$ and $\{a_1, a_2\} \subset [0,1]$ with $a_1 + a_2 = 1$, we now proceed by induction. Suppose $k > 2$ and suppose $a_1 x_1 \oplus \cdots \oplus a_{k-1} x_{k-1}$ has been defined, regardless of order, for all sets of $k - 1$ points of X and all $\{a_1, \cdots, a_{k-1}\} \subset [0,1]$ satisfying $\sum_{i=1}^{k-1} a_i = 1$. Now consider a k-tuple: $\{x_1, x_2, \cdots, x_k\} \subset X$ and suppose $\{a_1, \cdots, a_k\} \subset [0,1]$ satisfies $\sum_{i=1}^{k} a_i = 1$. By the inductive assumption we may further assume that $\{a_1, \cdots, a_k\} \subset (0,1)$. Now set

$$x_1^1 = a_1 x_1 \oplus (1 - a_1) \left(\frac{a_2}{1 - a_1} x_2 \oplus \frac{a_3}{1 - a_1} x_3 \oplus \cdots \oplus \frac{a_k}{1 - a_1} x_k \right)$$

$$x_2^1 = a_2 x_2 \oplus (1 - a_2) \left(\frac{a_1}{1 - a_2} x_1 \oplus \frac{a_3}{1 - a_2} x_3 \oplus \cdots \oplus \frac{a_k}{1 - a_2} x_k \right)$$

$$x_3^1 = a_3 x_3 \oplus (1 - a_3) \left(\frac{a_1}{1 - a_3} x_1 \oplus \frac{a_2}{1 - a_3} x_2 \oplus \cdots \oplus \frac{a_k}{1 - a_3} x_k \right)$$

$$\vdots$$

$$x_k^1 = a_k x_k \oplus (1 - a_k) \left(\frac{a_1}{1 - a_k} x_1 \oplus \frac{a_2}{1 - a_k} x_2 \oplus \cdots \oplus \frac{a_{k-1}}{1 - a_k} x_{k-1} \right)$$

In general, let

$$x_1^n = a_1 x_1^{n-1} \oplus (1 - a_1) \left(\frac{a_2}{1 - a_1} x_2^{n-1} \oplus \frac{a_3}{1 - a_1} x_3^{n-1} \oplus \cdots \oplus \frac{a_k}{1 - a_1} x_k^{n-1} \right)$$

$$x_2^n = a_2 x_2^{n-1} \oplus (1 - a_2) \left(\frac{a_1}{1 - a_2} x_1^{n-1} \oplus \frac{a_3}{1 - a_2} x_3^{n-1} \oplus \cdots \oplus \frac{a_k}{1 - a_2} x_k^{n-1} \right)$$

$$x_3^n = a_3 x_3^{n-1} \oplus (1 - a_3) \left(\frac{a_1}{1 - a_3} x_1^{n-1} \oplus \frac{a_2}{1 - a_3} x_2^{n-1} \oplus \cdots \oplus \frac{a_k}{1 - a_3} x_k^{n-1} \right)$$

$$\vdots$$

$$x_k^n = a_k x_k^{n-1} \oplus (1 - a_k) \left(\frac{a_1}{1 - a_k} x_1^{n-1} \oplus \frac{a_2}{1 - a_k} x_2^{n-1} \oplus \cdots \oplus \frac{a_{k-1}}{1 - a_k} x_{k-1}^{n-1} \right)$$

We now estimate $d\left(x_i^n, x_j^n \right)$, $i < j$. By iterated use of (i) we obtain

$$d\left(x_i^n, x_j^n \right) \leq \sum_{i=1}^{k} a_i d\left(x_i^{n-1}, x_j^n \right)$$

$$\leq \sum_{i=1}^{k} a_i \sum_{j=1}^{k} a_j d\left(x_i^{n-1}, x_j^{n-1} \right)$$

$$= \sum_{i,j=1}^{k} a_i a_j d\left(x_i^{n-1}, x_j^{n-1} \right)$$

$$\leq 2 \left[\sum_{i,j=1(i<j)}^{k} a_i a_j \right] diam\left(\{ x_1^{n-1}, x_2^{n-1}, x_3^{n-1}, \cdots, x_k^{n-1} \} \right).$$

In general for $i, j \in \{1, \cdots, k\}$, $i < j$,

$$d\left(x_i^n, x_j^n \right) \leq 2 \left[\sum_{i,j=1(i<j)}^{k} a_i a_j \right] diam\left(\{ x_1^{n-1}, x_2^{n-1}, x_3^{n-1}, \cdots, x_k^{n-1} \} \right),$$

and we conclude

$$diam\left(\{x_1^n, x_2^n, x_3^n, \cdots, x_k^n\}\right)$$

$$\leq\ 2\left[\sum_{i,j=1(i<j)}^{k} a_i a_j\right] diam\left(\{x_1^{n-1}, x_2^{n-1}, x_3^{n-1}, \cdots, x_k^{n-1}\}\right)$$

It remains to show that if $\{a_1, a_2, \cdots, a_n\} \subset (0, 1)$ and $\sum_{i=1}^{k} a_i = 1$, then

$$2\sum_{i,j=1(i<j)}^{k} a_i a_j < 1.$$

However

$$2\sum_{i,j=1(i<j)}^{k} a_i a_j\ =\ a_1\left(\sum_{j=2}^{k} a_j\right) + a_2\left(\sum_{j=1,j\neq 2}^{k} a_j\right) + \cdots + a_k\left(\sum_{j=1}^{k-1} a_j\right)$$

$$=\ a_1\left(1 - a_1\right) + a_2\left(1 - a_2\right) + \cdots + a_k\left(1 - a_k\right)$$

$$=\ \sum_{i=1}^{k} a_i - \sum_{i=1}^{k} a_i^2 = 1 - \sum_{i=1}^{k} a_i^2 < 1.$$

Letting

$$\delta = 2\sum_{i,j=1(i<j)}^{k} a_i a_j,$$

we now have

$$diam\left(\{x_1^n, x_2^n, x_3^n, \cdots, x_k^n\}\right) \leq \delta diam\left(\{x_1^{n-1}, x_2^{n-1}, x_3^{n-1}, \cdots, x_k^{n-1}\}\right)$$

with $\delta < 1$. It follows that

(6.1) $$diam\left(\{x_1^n, x_2^n, x_3^n, \cdots, x_k^n\}\right) \leq \delta^n diam\left(\{x_1, x_2, x_3, \cdots, x_k\}\right).$$

Now let $\overline{conv}\,(F)$ denote the closed convex hull of a subset $F \subset X$ in the usual sense. Thus $\overline{conv}\,(F)$ denotes the closure of the set

(6.2) $$conv\,(F) = \bigcup_{n=0}^{\infty} F_n,$$

where $F_0 = F$ and for $n \geq 1$ the set F_n consists of all points in the space which lie on geodesics which have endpoints in F_{n-1}. With this definition it is clear via (i) that $diam\,(F) = diam\,(F_1) = diam\,(F_2) = \cdots = diam\,(conv\,(F))$.

By construction, the set $\{x_1^n, x_2^n, x_3^n, \cdots, x_k^n\}$ lies in the convex hull of the set $\{x_1^{n-1}, x_2^{n-1}, x_3^{n-1}, \cdots, x_k^{n-1}\}$; thus

$$conv\,\{x_1^n, x_2^n, x_3^n, \cdots, x_k^n\} \subset conv\,\{x_1^{n-1}, x_2^{n-1}, x_3^{n-1}, \cdots, x_k^{n-1}\}.$$

Now, from inequality (6.1), we conclude that

$$diam\,(\overline{conv}\,\{x_1^n, x_2^n, x_3^n, \cdots, x_k^n\}) \leq \delta^n diam\,(\overline{conv}\,\{x_1, x_2, x_3, \cdots, x_k\}).$$

We can now apply Cantor's intersection theorem to the closures of the descending sequence of sets

$$\{conv\,\{x_1^n, x_2^n, x_3^n, \cdots, x_k^n\}\}_{n=1}^{\infty}$$

and conclude that for $1 \leq j \leq k$, each of the sequences $\{x_j^n\}_{n=1}^{\infty}$ is a Cauchy sequence, and all of these sequences converge to a common limit, which we denote $a_1 x_1 \oplus \cdots \oplus a_k x_k$.

As in the approach of [122], with this definition we have the following estimate: If $x, x_1, \cdots, x_n \in X$, then

$$d\,(x, a_1 x_1 \oplus \cdots \oplus a_k x_k) \leq \sum_{i=1}^{k} a_i d\,(x, x_i)\,.$$

If $a_i \equiv \dfrac{1}{k}$, then we have another definition of the mean point (or "barycenter") $\dfrac{x_1 \oplus \cdots \oplus x_k}{k}$ analogous to the one given in [98]. In this case,

$$2 \sum_{i,j=1(i<j)}^{k} a_i a_j = \frac{k-1}{k}$$

and for each $x \in X$,

$$d\left(x, \frac{x_1 \oplus \cdots \oplus x_k}{k}\right) \leq \frac{1}{k} \sum_{i=1}^{k} d\,(x, x_i)\,.$$

REMARK 6.1. *If X is a closed subset of a strictly convex Banach space, then the iterative process described above for defining the convex combination terminates at the first step. It is also known that X is isometric to a convex subset of a normed space if and only if affine functions on X separate points (see Theorem 1.1 in [94]).*

REMARK 6.2. *We will use the notation $co\,(F)$ to denote the collection of all convex combinations of finite subsets of F as defined above. All that is clear at this point is that $co\,(F)$ is contained in $conv\,(F)$. It is probably asking too much to expect the two sets to coincide. A third approach might be to set $F_0 = F$ and for $n \geq 1$, set $F_n = co\,(F_{n-1})$. It is now possible to define a new concept of "convex hull" of F by taking the union of the sets $co\,(F_n)$. In general this "convex hull" lies between $co\,(F)$ and $conv\,(F)$.*

CHAPTER 7

Length Spaces and Local Contractions

In general, a *path* in a metric space (X, d) is a continuous image of the unit interval $I = [0, 1] \subset \mathbb{R}$. If $S \equiv f(I)$ is a path, then its *length* is defined as

$$\ell(S) = \sup_{(x_i)} \sum_{i=0}^{N-1} d(f(x_i), f(x_{i+1}))$$

where $(x_i) := (0 = x_0 < x_1 < \cdots < x_N = 1)$ is any partition of $[0, 1]$. If $\ell(S) < \infty$, then the path is said to be *rectifiable*.

A metric space (X, d) is said to be a *length space* if the distance between each two points x, y of X is the infimum of the lengths of all rectifiable paths joining them. In this case, d is said to be a *length metric* (also known as *inner metric* or *intrinsic metric*).

A length space X is called a *geodesic space* if there is a path S joining each two points $x, y \in X$ for which $\ell(S) = d(x, y)$. Such a path is often called a *metric segment* (or a *geodesic*, as in the previous chapter) with endpoints x and y. There is a simple criterion which assures the existence of metric segments.

Another criterion is given in [**158**]. (Recall that a Hausdorff topological space X is said to be *locally compact* if each point has a neighborhood that is contained in a compact subspace of X.)

THEOREM 7.1. *If X is a complete metric space, locally compact at all except possibly one of its points, and any pair of points has a path of finite length joining them, then any pair of points has a shortest path joining them.*

There is an analog of Menger's criterion for length spaces.

DEFINITION 7.1 ([**93**]). *A metric space (X, d) is said to satisfy property* (A) *if given any two points $x, y \in X$, any two numbers $b, c \geq 0$ such that $b + c = d(x, y)$, and any $\varepsilon > 0$,*

(A) $$B(x; b + \varepsilon) \cap B(y; c + \varepsilon) \neq \emptyset.$$

The proof of Theorem 1 of [**93**] yields the following fact.

THEOREM 7.2. *If a complete metric space (X, d) satisfies property* (A) , *then each two points of X can be joined by a rectifiable path. (Thus X has an intrinsic metric.)*

© Springer International Publishing Switzerland 2014
W. Kirk, N. Shahzad, *Fixed Point Theory in Distance Spaces*,
DOI 10.1007/978-3-319-10927-5_7

The following is also known.

THEOREM 7.3. *Let K be a bounded convex subset of a Busemann convex space and let $f : K \to K$ be nonexpansive. Then* $\inf \{d(x, f(x)) : x \in K\} = 0$.

The question of whether there is an analog of this result for length spaces is complicated by the fact that it is not clear how to define Busemann convexity for length spaces. One way to circumvent this difficulty is by passing to a metric space ultrapower of the underlying space. There are many ways to do this. If (X, d) is a complete metric space, then X can be isometrically embedded in a Banach space E. It is now possible to identify X with its image in E. (The fact that one can do this is a classical result. In fact, E can be taken to be the space of all real valued continuous functions defined on X.) Now let $\tilde{E}$ denote the Banach space ultrapower of E relative to some nontrivial ultrafilter $\mathcal{U}$ over $\mathbb{N}$ in the usual sense (see, e.g., [**4**] for details). Thus the elements of $\tilde{E}$ are equivalence classes of bounded sequences $\tilde{x} := [(x_n)]$ in E, where $(u_n) \in [(x_n)]$ if and only if $\lim_{\mathcal{U}} \|u_n - x_n\| = 0$. Next take

$$\tilde{X} := \left\{ \tilde{x} = [(x_n)] \in \tilde{E} : x_n \in X \text{ for each } n \in \mathbb{N} \right\}.$$

Then for $\tilde{x}, \tilde{y} \in \tilde{X}$, set $\tilde{\rho}(\tilde{x}, \tilde{y}) = \lim_{\mathcal{U}} \|x_n - y_n\| = \lim_{\mathcal{U}} d(x_n, y_n)$. One can now say that a length space X is Busemann convex if and only if some ultrapower $\left(\tilde{X}, \tilde{\rho} \right)$ of X is Busemann convex in the usual sense.

THEOREM 7.4. *A complete metric space (X, d) is a length space if and only if every nontrivial ultrapower $\tilde{X}$ of X is a geodesic space.*

PROOF. Let $p, q \in X$ and $\alpha = (1/2)\rho(p, q)$. Let $\{\varepsilon_n\} \subset (0, \infty)$ with $\varepsilon_n \to 0$. The fact that X is a length space assures the existence of a sequence $\{m_n\} \subset B(p; \alpha + \varepsilon_n) \cap B(q; \alpha + \varepsilon_n)$. If $\tilde{m} = [(m_n)]$, then $\tilde{\rho}(\tilde{p}, \tilde{m}) = \tilde{\rho}(\tilde{q}, \tilde{m}) = (1/2)\tilde{\rho}(\tilde{p}, \tilde{q})$. Since $\tilde{X}$ is complete, X is a geodesic space by the criterion of Menger. On the other hand, if $\tilde{X}$ is a geodesic space, then it is easy to verify that X satisfies Property (A); hence, X is a length space by Theorem 7.2. $\qquad\square$

THEOREM 7.5. *Let K be a bounded Busemann convex length space and let $f : K \to K$ be nonexpansive. Then* $\inf \{d(x, f(x)) : x \in K\} = 0$.

PROOF. By assumption there is a nontrivial ultrapower $\tilde{X}$ of X that is Busemann convex. Define $\tilde{f} : \tilde{X} \to \tilde{X}$ by setting $\tilde{f}(\tilde{x}) = [f(x)] = \widetilde{f(x)}$. Then $\tilde{f}$ is also nonexpansive, so $\inf \left\{ \tilde{d}\left(\tilde{x}, \tilde{f}(\tilde{x}) \right) : \tilde{x} \in \tilde{X} \right\} = 0$. Thus given $\varepsilon > 0$ there exists $\tilde{x} \in \tilde{X}$ such that $\tilde{d}\left(\tilde{x}, \tilde{f}(\tilde{x}) \right) = \lim_{\mathcal{U}} d(x_n, f(x_n)) < \varepsilon$. This implies the existence of $x \in X$ for which $d(x, f(x)) < \varepsilon$. $\qquad\square$

(In the statement of the preceding result in [**121**] the boundedness assumption is inadvertently omitted.)

It is also shown in [158] that if (X, d) is a complete metric space which has the property that each two of its points can be joined by a rectifiable path, and if ℓ is the length metric on X, then (X, ℓ) is also complete. However this latter fact was proved earlier by Hu and Kirk [97]. It is implicit in the proof of the following theorem and stated as a corollary in [97]. (It is not true that if X is compact, then (X, ℓ) is compact—consider the radial metric on an appropriate subset of the unit disc in $\mathbb{R}^2$.)

DEFINITION 7.2. A mapping f defined on a metric space (X, d) is said to be a *local radial contraction* if there exists $k \in (0, 1)$ such that $d(f(x), f(u)) \leq kd(x, u)$ for u in some neighborhood U_x of x. (It follows that any local radial contraction is continuous.)

THEOREM 7.6. *Let (X, d) be a complete metric space and $f : X \to X$ a local radial contraction. Suppose for some $x_0 \in X$ the points x_0 and $f(x_0)$ are joined by a path of finite length. Then the sequence $\{f^n(x_0)\}$ converges to a fixed point of f.*

Rakotch proved the above theorem in [182] under the slightly stronger assumption that f is *locally contractive* in the sense that there exists $k \in (0, 1)$ such that each point of $x \in X$ has a neighborhood U_x such that $d(f(u), f(v)) \leq kd(u, v)$ for all $u, v \in U_x$.

The original proof of Theorem 7.6 in [97] was based on the following claim of Holmes in [96].

PROPOSITION 7.1 ([96]). *Let (X, d) be a compact metric space and suppose $f : X \to X$ is a local radial contraction. Then there exist numbers $k \in (0, 1)$ and $\beta > 0$ such that $d(f(x), f(y)) \leq kd(x, y)$ for all $x, y \in X$ such that $d(x, y) \leq \beta$.*

However G. Jungck has given an example in [103] which shows that this proposition is false. At the same time, Jungck has shown that the following is true.

PROPOSITION 7.2 ([103]). *Let (X, d) be a metric space and $g : X \to X$ a local radial contraction with contraction constant $k \in (0, 1)$. If $\alpha : [0, 1] \to X$ is a path of finite length $\ell(\alpha)$, then $g(\alpha)$ is also a path of finite length. Moreover $\ell(g(\alpha)) \leq k\ell(\alpha)$.*

Using Jungck's proposition the proof given in [97] carries over without essential change. We give the details.

PROOF OF THEOREM 7.6. Consider the space $\check{X}$ consisting of all those points of X that can be joined to x_0 by a rectifiable path and assign the length metric ℓ to $\check{X}$. We complete the proof by showing:

  (1) $f : \check{X} \to \check{X}$;
  (2) f is a contraction mapping on $(\check{X}, \ell)$;
  (3) $(\check{X}, \ell)$ is complete.

The conclusion will then follow from Banach's Theorem, i.e., f has a unique fixed point in $\check{X}$.

Let $y \in \check{X}$ and let α be a rectifiable path joining x_0 and y. By Proposition 7.2 the restriction of f to α maps α into a path β joining $f(x_0)$ and $f(y)$ with the property that $\ell(\beta) \leq k\ell(\alpha)$. By assumption there is a rectifiable path γ joining x_0 and $f(x_0)$. Thus $\beta \cup \gamma$ is a rectifiable path joining x_0 and $f(y)$. This proves (1). (2) is also a consequence of Proposition 7.1.

At this point it is possible to complete the proof by observing that $\{f^n(x_0)\}$ is a Cauchy sequence in $(\check{X}, \ell)$. Since $d(f^n(x_0), f^m(x_0)) \leq \ell(f^n(x_0), f^m(x_0))$, it follows that $\{f^n(x_0)\}$ is also a Cauchy sequence in (X, d). By completeness of (X, d), there exists $x^* \in X$ such that $\lim_{n \to \infty} d(f^n(x_0), x^*) = 0$, and since f is continuous, $f(x^*) = x^*$.

We now turn to (3). Let $\{x_n\}$ be a Cauchy sequence in $(\check{X}, \ell)$. Since $\ell(x, y) \geq d(x, y)$ it follows that $\{x_n\}$ is also a Cauchy sequence in (X, d); hence, there exists $x \in X$ such that $d(x_n, x) \to 0$. We complete the proof by showing that $x \in \check{X}$ and $\ell(x_n, x) \to 0$.

Let $\{\varepsilon_i\}$ be a sequence of positive numbers for which $\sum_{i=1}^{\infty} \varepsilon_i < \infty$. Since $\{x_n\}$ is Cauchy in $(\check{X}, \ell)$, there exist positive integers $\{N_i\}$ such that $m, n \geq N_i \Rightarrow \ell(x_n, x_m) \leq \varepsilon_i$. It is now possible to choose a subsequence $\{\bar{x}_n\}$ of $\{x_n\}$ such that $\ell(\bar{x}_n, \bar{x}_{n+1}) < \varepsilon_n$, $n = 1, 2, \cdots$. For each n there is a path $\alpha_n : \left[\frac{1}{n+1}, \frac{1}{n}\right] \to (\check{X}, d)$ joining $\bar{x}_n$ and $\bar{x}_{n+1}$ with length less than ε_n. Define $\alpha : [0, 1] \to \check{X}$ by taking

$$\alpha(t) = \begin{cases} \alpha_n(t) & \text{if } t \in \left[\frac{1}{n+1}, \frac{1}{n}\right], \\ x & \text{if } t = 0. \end{cases}$$

Clearly α is continuous on $(0, 1]$. To see that α is continuous at 0, let $t_i \downarrow 0$. Then given any $N \in \mathbb{N}$ and i sufficiently large, $t_i \in \left[\frac{1}{n+1}, \frac{1}{n}\right]$ for some $n \geq N$. It follows that

$$d(\alpha(t_i), x) \leq d\left(\alpha\left(\frac{1}{n+1}\right), x\right) + d\left(\alpha\left(\frac{1}{n+1}\right), \alpha(t_i)\right)$$
$$\leq d(\bar{x}_n, x) + \varepsilon_n.$$

From this it follows that $\lim_{i \to \infty} d(\alpha(t_i), x) = 0$. This proves continuity of α at 0. Also $\ell(\alpha) \leq \sum_{i=1}^{\infty} \varepsilon_i$.

Notice that

$$\ell(\bar{x}_n, x) \leq \sum_{i=n}^{\infty} \ell(\bar{x}_i, \bar{x}_{i+1}) \leq \sum_{i=n}^{\infty} \varepsilon_i.$$

Therefore $\lim_{n \to \infty} \ell(\bar{x}_n, x) = 0$. Since $\{\bar{x}_n\}$ is a subsequence of the Cauchy sequence $\{x_n\}$ in $(\check{X}, \ell)$, it follows that $\lim_{n \to \infty} \ell(x_n, x) = 0$, and we are finished. $\square$

Implicit in the above proof is the following fact.

THEOREM 7.7. *Let (X, d) be a complete metric space, and suppose each two points of X can be joined by a rectifiable path. Then (X, ℓ) is also complete, where ℓ is the length metric on X induced by d. Consequently every local radial contraction $f : X \to X$ has a unique fixed point x^*, and moreover $\lim_{n\to\infty} f^n(x) = x^*$ for each $x \in X$.*

An example is given in [97] (see Example 7.2 in the next section) shows that Theorem 7.6 is false if x_0 and $g(x_0)$ are merely assumed to be joined by an arbitrary path rather than a rectifiable path. Also an earlier example in [182] shows that the fixed point in Theorem 7.6 need not be unique, even if the space is connected.

PROPOSITION 7.3 ([10]). *A connected open subset of a geodesic space has a path metric.*

PROOF. Let U be a connected open subset of a geodesic space and let $x \in U$. Let

$$U_0 = \{y \in U : x \text{ and } y \text{ can be joined by a rectifiable path}\}.$$

If $y \in U$, then some open ball centered at y also lies in U, and any point in this ball is clearly in U_0. So U_0 is an open subset of U. Suppose U_0 is a proper subset of U and let $u \in U \backslash U_0$. Then some open ball centered at u lies in U, and this ball must necessarily lie in $U \backslash U_0$. This would mean that U is the union of two disjoint open sets, which is clearly impossible because U is connected. Hence $U_0 = U$. □

In the following theorem, $\overline{U}$ denotes the closure of U.

THEOREM 7.8 ([10]). *Let U be a connected open subset of a complete geodesic space (X, d), suppose $f : U \to U$ is a local radial contraction, and suppose f can be extended to a continuous mapping $\overline{f} : \overline{U} \to \overline{U}$. Then $\overline{f}$ has a fixed point in $\overline{U}$, and moreover $\{f^n(x)\}$ converges to x^* for each $x \in U$.*

PROOF. Let ℓ be the path metric on U. In view of proof of Theorem 7.6 f is a contraction mapping on (U, ℓ). Let $x \in U$. By a standard argument $\{f^n(x)\}$ is a Cauchy sequence in (U, ℓ). This in turn implies that $\{f^n(x)\}$ is a Cauchy sequence in (U, d). Hence $\{f^n(x)\}$ converges to some point $x^* \in \overline{U}$. Since $\overline{f}$ is continuous, we conclude $\overline{f}(x^*) = x^*$. Moreover if k is the contraction constant for f and if $y \in U$, then $\ell(f^n(x), f^n(y)) \leq k^n \ell(x, y)$. It follows that $\lim_{n\to\infty} \ell(f^n(x), f^n(y)) = 0$ and so $\{f^n(y)\}$ converges to x^*. Finally, if for some $x \in U$ the segment $[x, f(x)]$ lies in U, then we have the estimate

$$d(f^n(x), x^*) \leq \ell(f^n(x), x^*) \leq \frac{k^n}{1-k} \ell(x, f(x)) = \frac{k^n}{1-k} d(x, f(x)).$$

□

EXAMPLE 7.1. *At this point it is probably natural to wonder whether the closure of a connected open subset of a Banach space always has a path metric. The answer is no. An example can be given in $\mathbb{R}^2$. Let $\varepsilon \in (0, 1/2)$ and R the open rectangle with vertices $(0,0)$, $(0, 1 + \varepsilon)$, $(1, 1 + \varepsilon)$, $(1, 0)$. Delete the closed strip centered on the segment joining $(1/2, 0)$ to $(1/2, 1)$ of width $1/6$. Then delete the closed strip centered on the segment joining $(1/3, 1 + \varepsilon)$ to $(1/3, \varepsilon)$ of width $1/12$. In general delete the closed strip centered on the segment joining $(1/2n, 0)$ and $(1/2n, 1)$ of width $1/[2n(2n+1)]$ and delete the closed strip centered on the segment joining $(1/(2n+1), 1 + \varepsilon)$ and $(1/(2n+1), \varepsilon)$ of width $1/[(2n+1)(2n+2)]$. Now let U be the points of R remaining after the closed strips have been deleted. Clearly U is a connected open set in $\mathbb{R}^2$. However the point $(0, 1/2)$ is in the closure of U, but no path of finite length can join any point of U to $(0, 1/2)$.*

THEOREM 7.9 ([**10**]). *Let D be the closure of a connected open set in a Banach space, and suppose D is rectifiably pathwise connected. Then any local radial contraction $f : D \to D$ has a unique fixed point.*

THEOREM 7.10 ([**10**]). *Let U be a connected open set in a Banach space X, and suppose the intersection of every line in X with U consists of at most finitely many open intervals. Then $\overline{U}$ is rectifiably pathwise connected. Consequently every local radial contraction $f : \overline{U} \to \overline{U}$ has a unique fixed point.*

PROOF. Let $x, y \in \overline{U}$. For $z \in U$, the line $L(z, x)$ passing through z and x intersects U in a finite number of open intervals. Consequently there is a metric segment $[u, x]$ lying on this line with $[u, x] \subset \overline{U}$ and $u \in U$. Similarly there is a metric segment $[v, y] \subset \overline{U}$ with $v \in U$. By Proposition 7.3 there is a rectifiable path α joining u and v. It follows that $\alpha \cup [u, x] \cup [v, y]$ is a rectifiable path joining x and y. The result now follows from Theorem 7.9. □

It is not difficult to think of very elaborate examples of open sets in Banach spaces which satisfy the criteria of Theorem 7.10. In fact a more general formulation is true.

THEOREM 7.11 ([**10**]). *Let U be a connected open set in a Banach space, and suppose for each $x \in \overline{U}$, there exists $z \in U$ such that the interval (x, z) lies in U. Then $\overline{U}$ is rectifiably pathwise connected.*

The following result is Theorem 1 in Holmes [**96**].

THEOREM 7.12. *Let (X, d) be a connected and locally connected metric space and let f be a homeomorphism of X onto X which is a local radial contraction. Then there is a metric ρ on X, topologically equivalent to d, such that f is a contraction on (X, ρ).*

Holmes also asserts in a corollary in [**96**] that completeness of (X, ρ) follows from completeness of (X, d). This in turn would imply that f has a fixed point if (X, d) is complete. However, in view of the example given in the next section (Example 7.2), either the theorem is false or the assertion of the corollary is false.

The following lemma is central to the proof of Theorem 7.12.

LEMMA 7.1 ([96]). *If f^n is a contraction on (X, d) and if f is continuous, then for each k, $0 < k < 1$, there exists a metric ρ on X, equivalent to d, such that f is a k-contraction on (X, ρ).*

The proof applies the following theorem of P. Meyers [154]. (Holmes neglects to mention that f is continuous, but it is obvious from his proof that this assumption is necessary.)

THEOREM 7.13. *Let (X, d) be a metric space. Suppose $f : X \to X$ is continuous and satisfies:*
   (i) *there exists $x^* \in X$ such that $f(x^*) = x^*$;*
   (ii) *$f^n(x) \to x^*$ as $n \to \infty$ for all $x \in X$;*
   (iii) *there is an open neighborhood U of x^* such that $f^n(U) \to \{x^*\}$ as $n \to \infty$ (i.e., for any open neighborhood V of x^* there is an $n(V) > 0$ such that $f^n(U) \subset V$ for all $n \geq n(V)$).*
Then for each $k \in (0,1)$ there is a metric ρ on X such that f is a k-contraction on (X, ρ). Moreover if (X, d) is complete, then so is (X, ρ).

PROOF OF LEMMA 7.1. We proceed to show that (i), (ii), (iii) hold under the assumptions of Lemma 7.1.

If the contraction mapping f^n does not have a fixed point, then by the Banach Contraction Principle we may adjoin a point x^* to X which will be the unique fixed point of f^n. In either case $f^i(x) \to x^*$ as $i \to \infty$ for each $x \in X$ and conditions (i) and (ii) are fulfilled. To see this, observe that $f^n(f(x^*)) = f(f^n(x^*)) = f(x^*)$. Thus $f(x^*)$ is a fixed point of f^n. Since the fixed point of f^n is unique, we must have $f(x^*) = x^*$. So (i) is true. But why is (ii) true? Because $i \in \mathbb{N} \Rightarrow i = nj + t$ for some $0 \leq t \leq n - 1$, so for $x \in X$

$$f^i(x) = f^{nj+t}(x) = f^{nj}(f^t(x)) \to x^* \text{ as } i \to \infty.$$

Note that f^{nj} converges to x^* uniformly on the finite set

$$S := \{x, f(x), \cdots, f^{n-1}(x)\}.$$

For (iii) set $V = B(x^*; 1)$ and let λ be the contraction constant of f^n. Then if $v \in V$ and $j \in \mathbb{N}$,

$$d\left(f^{nj}(v), f^{nj}(x^*)\right) \leq \lambda^j d(v, x^*)$$

so $f^{nj}(V) \subset B(x^*; \lambda^j)$. Set $U = \cap_{i=0}^{n-1} f^{-i}(V)$. Now notice that **since f is continuous**, U is a neighborhood of x^*, and, if $0 \leq t < n$,

$$f^{nj+t}(U) \subset f^{nj}(V) \subset B(x^*; \lambda^j)$$

and condition (iii) is fulfilled.   □

REMARK 7.1. *In* [201] *it is shown that if* (X,d) *is a metric space and if* $f : X \to X$ *is a contraction with constant* λ, *then for any* λ *such that* $\lambda^{1/n} < k < 1$ *there is a metric* ρ *on* X *such that* f *is a* k-*contraction on* (X,ρ). *Moreover if* f *is uniformly continuous on* (X,d) *and if* d *is complete, then so is* ρ.

Implicit in Lemma 7.1 is the following fact.

THEOREM 7.14. *Let* X *be a complete metric space and* $f : X \to X$ *a mapping for which* f^N *is a contraction for some* $N \in \mathbb{N}$. *Then* f *has a unique fixed point* x^* *and* $\lim_{n\to\infty} f^n(x) = x^*$ *for each* $x \in X$.

This is actually a special case of a more general topological result which has been known for some time. We prove the metric case here.

THEOREM 7.15. *Let* X *be a metric space, let* $x^* \in X$, *and let* $f : X \to X$ *be a mapping for which* $g := f^N$ *satisfies* $\lim_{n\to\infty} g^n(x) = x^*$ *for each* $x \in X$. *Then* $\lim_{n\to\infty} f^n(x) = x^*$ *for each* $x \in X$.

PROOF. Let $\varepsilon > 0$ and let $x \in X$. By assumption there exists $N_1 \in \mathbb{N}$ such that $j \geq N_1 \Rightarrow d\left(f^{jN}(x), x^*\right) \leq \varepsilon$. Similarly there exists $N_2 \in \mathbb{N}$ such that $j \geq N_2 \Rightarrow d\left(f^{jN}(f(x)), x^*\right) = d\left(f^{jN+1}(x), x^*\right) \leq \varepsilon$. In general, for each $i \in \{0, \cdots, n-1\}$ there exists $N_i \in \mathbb{N}$ such that $j \geq N_i \Rightarrow$

$$d\left(f^{jN}\left(f^i(x)\right), x^*\right) = d\left(f^{jN+i}(x), x^*\right) \leq \varepsilon.$$

Finally, there exists $N \in \mathbb{N}$ such that $n \geq N \Rightarrow n = jN + i$ for some $j \geq \max\{N_1, \cdots, N_{n-1}\}$ and $i \in \{0, \cdots, n-1\}$. Thus $n \geq N \Rightarrow d\left(f^n(x), x^*\right) \leq \varepsilon$. $\qquad\square$

COROLLARY 7.1. *Let* X *be a complete metric space and* $f : X \to X$ *a mapping for which* f^N *is an asymptotic contraction for some* $N \in \mathbb{N}$. *Then* f *has a unique fixed point* x^* *and* $\lim_{n\to\infty} f^n(x) = x^*$ *for each* $x \in X$.

COROLLARY 7.2. *Let* X *be a complete metric space for which each two points can be joined by a rectifiable path, and suppose* $f : X \to X$ *is a mapping for which* f^N *is a local radial contraction for some* $N \in \mathbb{N}$. *Then* f *has a unique fixed point* x^*, *and* $\lim_{n\to\infty} f^n(x) = x^*$ *for each* $x \in X$.

7.1. Local Contractions and Metric Transforms

We now turn to a special case of a classical concept due to L.M. Blumenthal (see [28, p. 130]). We call strictly increasing concave function $\phi : \mathbb{R}^+ \to \mathbb{R}$ for which $\phi(0) = 0$ a *metric transform*. It is known (see Exercise 5 in [28, p. 26]) that if (X,d) is a metric space and if $\rho(x,y) = \phi(d(x,y))$ for each $x, y \in X$ for such a function ϕ, then (X,ρ) is also a metric space. Blumenthal had introduced this concept earlier in [27] to show that the metric transform $\phi(X)$ of any metric space X, where $\phi(t) = t^\alpha$, $0 < \alpha \leq \frac{1}{2}$, has the Euclidean four point property, i.e., each four points of $\phi(X)$ are isometric to a quadruple of points in three-dimensional Euclidean space.

We now give a simple condition in terms of metric transforms which implies that a mapping $f : X \to X$ is a local radial contraction. Notice that if ϕ is taken to be the identity mapping, the following result reduces to the definition of a local radial contraction. (This discussion is taken from [**128**].)

THEOREM 7.16. *Let (X, d) be a metric space and $f : X \to X$. Suppose there exist a metric transform ϕ and a number $k \in (0, 1)$ such that the following conditions hold:*

(a) *For each $x \in X$ there exists $\varepsilon_x > 0$ such that*

$$d(x, u) < \varepsilon_x \Rightarrow \phi(d(f(x), f(u))) \leq kd(x, u).$$

(b) *There exists $c \in (0, 1)$ such that for all $t > 0$ sufficiently small*

$$kt \leq \phi(ct).$$

Then f is a local radial contraction on (X, d).

In view of Theorem 7.7 we now have the following.

THEOREM 7.17. *Suppose, in addition to the assumptions in Theorem 7.16, X is complete and rectifiably pathwise connected. Then f has a unique fixed point x^*, and $\lim_{n \to \infty} f^n(x) = x^*$ for each $x \in X$.*

PROOF OF THEOREM 7.16. Let $x \in X$. Then if $d(x, u) < \varepsilon_x$,

$$\phi(d(f(x), f(u))) \leq kd(x, u).$$

Now suppose there exists $c \in (0, 1)$ such that for t sufficiently small,

$$kt \leq \phi(ct).$$

This implies there exists $\delta_x > 0$ with $\delta_x \leq \varepsilon_x$ such that $d(x, u) < \delta_x \Rightarrow$

$$\phi(d(f(x), f(u))) \leq kd(x, u) \leq \phi(cd(x, u)).$$

Since ϕ is strictly increasing, $d(x, u) < \delta_x \Rightarrow$

$$d(f(x), f(u)) \leq cd(x, u).$$

Therefore f is a local radial contraction on (X, d).   □

REMARK 7.2. *If condition (a) is changed to*

$$\phi(d(f(x), f(y))) \leq kd(x, y) \text{ for all } x, y \in X,$$

then f is a uniform local contraction on (X, d). This is because condition (b) now implies that there exists $\delta > 0$ such that $d(x, y) < \delta \Rightarrow$

$$\phi(d(f(x), f(y))) \leq kd(x, y) \leq \phi(cd(x, y)).$$

REMARK 7.3. *If $f : X \to X$ is onto and satisfies the following expansive type condition: there exists $k \in (0,1)$ such that*

$$d\left(f\left(x\right), f\left(y\right)\right) \geq k^{-1}\phi\left(d\left(x,y\right)\right) \ \text{for all } x, y \in X,$$

then f^{-1} is a uniform local contraction on (X, d). This is because f^{-1} exists and satisfies

$$\phi\left(d\left(f^{-1}\left(x\right), f^{-1}\left(y\right)\right)\right) \leq kd\left(x,y\right) \ \text{for all } x, y \in X.$$

Condition (b) of Theorem 7.16 might appear to be too restrictive. However we now list several examples of nontrivial metric transforms for which the condition holds.

(i) $\phi\left(t\right) = \dfrac{t}{1+t}$. Let $k \in (0,1)$ and select $c \in (k, 1)$. Then

$$kt \leq \phi\left(ct\right) \Leftrightarrow kt \leq \frac{ct}{1+ct} \Leftrightarrow$$

$$k \leq \frac{c}{1+ct} \Leftrightarrow t \leq \frac{c-k}{ck}.$$

Since $c > k$, condition (b) follows.

(ii) $\phi\left(t\right) = t^{\beta}$, for $\beta \in (0,1)$. Then for any $c, k \in (0,1)$

$$t \leq \frac{\phi\left(ct\right)}{k} \Leftrightarrow t \leq \frac{\left(ct\right)^{\beta}}{k},$$

and condition (b) holds for $t \leq 1$.

(iii) $\phi\left(t\right) = \sin\left(\dfrac{t}{1+t}\right)$. Let $k \in (0,1)$, and set $h\left(t\right) = \dfrac{t}{1+t}$. We know that if $c \in (k, 1)$ and if $t \leq \dfrac{c-k}{ck}$, then

$$kt \leq h\left(ct\right).$$

In particular, take $k' \in (k, 1)$, then choose $c \in (k', 1)$. The same argument as in (ii) shows that if $t \leq \dfrac{c-k'}{ck'}$, then

$$kt < k't \leq h\left(ct\right).$$

Thus if t is sufficiently small,

$$kt \leq \sin k't \leq \sin\left(h\left(ct\right)\right) = \phi\left(ct\right).$$

(iv) $\phi\left(t\right) = p\tan^{-1}t$ for fixed $p > 1$. Let $k \in (0,1)$. Then $kt \leq \phi\left(ct\right) \Leftrightarrow$ $\tan\left(\dfrac{kt}{p}\right) \leq ct$. Let $f\left(t\right) = ct - \tan\left(\dfrac{kt}{p}\right)$. Then $f\left(0\right) = 0$ and

$$f'\left(t\right) = c - \frac{k}{p}\sec^2\left(\frac{kt}{p}\right) > 0 \Leftrightarrow \sec^2\left(\frac{kt}{p}\right) < \frac{pc}{k}.$$ If $c \in (0,1)$ is chosen so that $\dfrac{pc}{k} > 1$, then $f'\left(t\right) > 0$ for $t > 0$ sufficiently small. This implies that $f\left(t\right) > 0$ for $t > 0$ sufficiently small, and this in turn implies that condition (b) holds.

(v) $\phi(t) = \ln(1 + t)$. Let $k \in (0,1)$ and select $c \in (k,1)$ Then $kt \leq \phi(ct) \Leftrightarrow e^{kt} \leq 1 + ct$. Let $f(t) = 1 + ct - e^{kt}$. Then $f(0) = 0$ and for $t > 0$, $f'(t) > 0 \Leftrightarrow e^{kt} < \dfrac{c}{k} \Leftrightarrow t < k^{-1} \ln\left(\dfrac{c}{k}\right)$. This is clearly true for $t > 0$ sufficiently small because $c \in (k,1)$.

Not every metric transform satisfies condition (b); $\phi(t) = \tan^{-1} t$ provides an example. On the other hand, Proposition 7.4 below shows that the collection of metric transforms which do satisfy condition (b) are indeed numerous and complex.

PROPOSITION 7.4. *Let $\mathfrak{M}$ denote the class of all metric transforms ϕ with the property that ϕ is twice differentiable, and let $\mathfrak{M}_1$ denote the subfamily of $\mathfrak{M}$ consisting of those $\phi \in \mathfrak{M}$ which satisfy the following condition: for any $k \in (0,1)$ there exists $c \in (0,1)$ such that for $t > 0$ sufficiently small,*

$$kt \leq \phi(ct).$$

Then both $\mathfrak{M}$ and $\mathfrak{M}_1$ are closed under functional composition.

PROOF. Let $\phi, \psi \in \mathfrak{M}$ and let $\varphi = \phi \circ \psi$. Then $\varphi(0) = \phi \circ \psi(0) = 0$. Also for any $t > 0$,

$$\varphi'(t) = \phi'(\psi(t)) \cdot \psi'(t) > 0$$

and

$$\varphi''(t) = \phi'(\psi(t)) \cdot \psi''(t) + \phi''(\psi(t)) \cdot [\psi'(t)]^2 < 0.$$

Therefore $\varphi \in \mathfrak{M}$.

Now suppose $\phi, \psi \in \mathfrak{M}_1$. Then there exists $c_1 \in (0,1)$ such that for $t > 0$ sufficiently small,

$$kt \leq \phi(c_1 t).$$

Also there exists $c \in (0,1)$ such that for $t > 0$ sufficiently small

$$c_1 t \leq \psi(ct).$$

Since ϕ is strictly increasing,

$$c_1 t \leq \psi(ct) \Leftrightarrow \phi(c_1 t) \leq \phi(\psi(ct)).$$

Therefore $kt \leq \varphi(ct)$ for $t > 0$ sufficiently small, so it follows that $\varphi \in \mathfrak{M}_1$.   □

The following example was given in [**97**]. It shows that Theorem 7.17 is false if the space is merely assumed to be pathwise connected rather than rectifiably pathwise connected. This illustrates another application of the idea of metric transforms.

EXAMPLE 7.2. *Let $(\beta_n)_{n=-\infty}^{\infty}$ be a strictly increasing doubly infinite sequence in $(0,1)$. For $x, y \in \mathbb{R}^+$, set*

$$(7.1) \qquad \rho(x,y) = \begin{cases} |x-y|^{\beta_n} & \text{if } x, y \in [n, n+1]; \\ |x-(n+1)|^{\beta_n} + (p-1) + |(n+p)-y|^{\beta_{n+p}} \\ \text{if } x \in [n, n+1], \ y \in [n+p, n+p+1], \ p \in \mathbb{N}. \end{cases}$$

We first observe that $(\mathbb{R}^+, \rho)$ is a metric space (see Proposition 7.5 below).

Now define $f : \mathbb{R}^+ \to \mathbb{R}^+$ by setting $f(x) = x + 1$. This mapping is a homeomorphism which is a local contraction for any $k \in (0,1)$. To see this, suppose $x, y \in [n, n+1]$. Then

$$\rho(f(x), f(y)) = |x-y|^{\beta_{n+1}} \le k |x-y|^{\beta_n} = k\rho(x,y)$$

if and only if $|x-y|^{\beta_{n+1}-\beta_n} \le k$. Since $\beta_{n+1} - \beta_n > 0$, this is always true if $|x-y|$ is sufficiently small; indeed

$$\rho(x,y) = |x-y|^{\beta_n} \le k^{\beta_n/(\beta_{n+1}-\beta_n)} \Leftrightarrow |x-y|^{\beta_{n+1}-\beta_n} \le k.$$

To deal with the case $x = n > 0$, merely take a neighborhood of x with radius less than $\min\left\{ k^{\beta_n/(\beta_{n+1}-\beta_n)}, k^{\beta_{n+1}/(\beta_{n+2}-\beta_{n+1})} \right\}$.

Notice that the mapping of the above example is even locally contractive in the sense of Rakotch [182], but it is fixed point free. We note also that the space $(\mathbb{R}^+, \rho)$ is topologically equivalent to $\mathbb{R}^+$ with its usual metric. In particular $(\mathbb{R}^+, \rho)$ is complete, connected, and locally connected.

The technique of the example is a special case of "gluing" of metric spaces (see, e.g., [36, p. 67]). Specifically, we use the following fact, which is a special case of Lemma 5.34 of [36].

PROPOSITION 7.5. *Suppose (M_1, d_1) and (M_2, d_2) are metric spaces with $M_1 \cap M_2 = \{u\}$. For $x, y \in X := M_1 \cup M_2$ set*

$$\rho(x,y) = d_i(x,y) \ \text{if } x, y \in M_i, \ i = 1, 2;$$
$$\rho(x,y) = d_1(x,u) + d_2(u,y) \ \text{if } x \in M_1, \ y \in M_2.$$

Then (X, ρ) is a metric space.

We now observe that for each $n \in \mathbb{Z}$ and $\beta_n \in (0,1)$, the metric transform $\phi_n(t) = t^{\beta_n}$ induces a metric on the interval $[n, n+1]$. The metric space $(\mathbb{R}^+, \rho)$ is obtained by simply "gluing" the consecutive intervals at their common endpoints and applying Proposition 7.5 inductively. This results in the metric defined by (7.1).

Remark. A theorem which appears to be a slight extension of Theorem 7.16 has recently been announced. A mapping $\varphi : \mathbb{R}^+ \to \mathbb{R}^+$ is said to be *metric preserving* if for all metric spaces (X, d), $\varphi \circ d$ is a metric on X. It is known that if φ is metric preserving, then $\varphi'(0)$ in the extended sense always exists (see [54] for a survey).

The following is the main result of [172].

THEOREM 7.18. *Let (X, d) be a metric space and let $f : X \to X$. Assume that there exist $k \in (0, 1)$ and a metric preserving function φ satisfying the following conditions:*

(a) For each $x \in X$ there exists $\varepsilon_x > 0$ such that for every $u \in X$

$$d(x, u) < \varepsilon_x \Rightarrow (\varphi \circ d)(f(x), f(u)) \leq kd(x, u).$$

(b) $\varphi'(0) > k$.

Then f is a local radial contraction.

CHAPTER 8

The G-Spaces of Busemann

Here we digress somewhat, although fixed point theory in geodesic spaces is an important underlying factor. A finitely compact (recall, this means bounded closed sets are compact) geodesically connected (metrically convex) metric space (R, d) which has the geodesic extension property (see Definition 9.3 below) and for which such extension is unique is called a G-space. This definition is due to Busemann [46]. Precisely, to every point $p \in R$ there corresponds a number $\rho_p > 0$ such that if $x, y \in U(p; \rho_p)$ (the open ball) with $x \neq y$, there exists a point $z \in R$ for which

$$d(x, y) + d(y, z) = d(x, z),$$

and moreover, the conditions $d(x, y) + d(y, z_1) = d(x, z_1)$, $d(x, y) + d(y, z_2) = d(x, z_2)$, and $d(y, z_1) = d(y, z_2) \Rightarrow z_1 = z_2$. A mapping ψ of a G-space $\tilde{R}$ onto a G-space R is called a *local isometry* if for every $\bar{p} \in \tilde{R}$, there exists a number $\eta_{\bar{p}} > 0$ such that ψ maps $U(\bar{p}; \eta_{\bar{p}})$ isometrically onto $U(p; \eta_p)$. When such a mapping exists the space $\tilde{R}$ is said to be a *covering space* of the G-space R and ψ a *covering map*. It is shown in [46] that every G-space has a *universal* (simply connected) covering space which is unique up to isometries (and which is, itself, a G-space). In particular, if $\tilde{R}$ is the universal covering space of R with $\Omega : \tilde{R} \to R$ a covering map, then the fundamental group of R may be realized as the group of those motions Ψ (surjective isometries) of $\tilde{R}$ onto $\tilde{R}$ for which $\Omega \circ \Psi = \Omega$.

An isometry of a G-space onto itself is called a *motion*. It is known that a local isometry of a noncompact G-space R onto itself is a motion if the fundamental group of R is not isomorphic with a proper subgroup of itself [46, p. 174]. Without this hypothesis on the fundamental group the assertion may or may not be true. It is true for a cylinder with a locally Euclidean metric but it is false for a cylinder with a locally hyperbolic metric. This leads to the problem (see [46, p. 405, (27)]) of finding conditions under which local isometries are motions, in particular conditions which apply to an ordinary cylinder. A response to this problem is given in [108]. To discuss this further we need some fundamental properties of local isometries.

Let ϕ denote a locally isometric mapping of a G-space $(\bar{R}, \bar{d})$ onto a G-space (R, d). The following properties of ϕ are found in Busemann [46, pp. 167–170].

W. Kirk, N. Shahzad, *Fixed Point Theory in Distance Spaces*,
DOI 10.1007/978-3-319-10927-5_8

(1) If $\bar{x}(\tau)$, $\alpha \leq \tau \leq \beta$, is a curve in $\bar{R}$ and if $\phi(\bar{x}(\tau)) = x(\tau)$ is a geodesic segment in R, then $\bar{x}(\tau)$ is a geodesic segment and $\bar{d}(\bar{x}(\alpha), \bar{x}(\beta)) = d(x(\alpha), x(\beta))$.

(2) For a given curve $x(\tau)$, $\alpha \leq \tau \leq \beta$, in R and a given point $\bar{a} \in \bar{R}$ such that $\phi(\bar{a}) = x(\alpha)$ there is exactly one curve $\bar{x}(\tau)$ in $\bar{R}$ such that $\phi(\bar{x}(\tau)) = x(\tau)$ with $\bar{x}(\alpha) = \bar{a}$.

(3) Given $p \in R$ there is a number $\rho(p) > 0$ such that if $\bar{p}_1, \bar{p}_2 \in \bar{R}$ satisfy $\bar{p}_1 \neq \bar{p}_2$ and $\phi(\bar{p}_1) = \phi(\bar{p}_2) = p$, then $\bar{d}(\bar{p}_1, \bar{p}_2) \geq 2\rho(p)$.

(4) The number of points of $\bar{R}$ that are mapped into a given point of R is at most countable, and is the same for different points of R.

(5) If ϕ is one-to-one, then ϕ is an isometry.

(6) If $a, b \in R$ and if $\phi(\bar{a}) = a$, there is exactly one point $\bar{b} \in \bar{R}$ such that $\phi(\bar{b}) = b$ and $\bar{d}(\bar{a}, \bar{b}) = d(a, b)$.

The following is immediate from the definition of a local isometry.

(7) If ϕ is a locally isometric mapping of R onto itself, then ϕ^n is also, $n = 1, 2, \cdots$.

THEOREM 8.1 ([108]). *A locally isometric mapping of a G-space onto itself which has a fixed point is a motion.*

PROOF. Let $\phi(p) = p$, and suppose ϕ is not a motion. Then by (5) ϕ is not one-to-one so by (4) there is a point $p_1 \in R$ with $p_1 \neq p$ such that $\phi(p_1) = p$. By (6) there is a point $p_2 \in R$ such that $\phi(p_2) = p_1$ and such that $d(p, p_2) = d(p, p_1)$. Proceeding by induction obtain a sequence $\{p_n\} \subset R$ such that $\phi(p_{n+1}) = p_n$ and $d(p, p_n) = d(p, p_1)$, $n = 1, 2, \cdots$.

If $n < m$, then $\phi^n(p_n) = p$ while $\phi^n(p_m) = p_{m-n}$. Since

$$d(p, p_{m-n}) = d(p, p_1) > 0,$$

we see that $p_{m-n} \neq p$. Therefore $\phi^n(p_n) \neq \phi^n(p_m)$ and it follows that $p_n \neq p_m$. By (7) ϕ^n, for each positive integer n, is a locally isometric mapping of R onto itself, so by (3) $d(p_i, p_j) \geq 2\rho(p)$ if $i \neq j$. Since the sequence $\{p_n\}$ is bounded, this contradicts the finite compactness of R. $\square$

THEOREM 8.2. *A locally isometric mapping ϕ of a G-space (R, d) onto itself is a motion if and only if there exists a motion ψ of R such that for some point $p \in R$, $\psi \circ \phi(p) = p$.*

PROOF. The necessity is trivial. The sufficiency is established by observing that $\psi \circ \phi$ is a locally isometric mapping with a fixed point p. Thus by Theorem 8.1 $\psi \circ \phi$ is a motion, and hence one-to-one. Therefore ϕ is one-to-one, and by (5) also a motion. $\square$

The group of motions of a G-space are said to be transitive if given any two points of the space there is a motion of the space that maps one into the other. Among two-dimensional G-spaces, it is known that the cylinder (and torus) with a Minkowskian metric has a transitive group of motions. Thus the following is an immediate consequence of Theorem 8.2.

THEOREM 8.3 ([**108**]). *If a G-space R has a transitive group of motions, then every locally isometric mapping of R onto itself is a motion.*

Other conditions are known to imply that locally isometric mappings are always motions. For example:

THEOREM 8.4 ([**109**]). *If ϕ is a locally isometric mapping of a G-space onto itself and if $\{\phi^n(p)\}$ is bounded, then ϕ is a motion.*

It is also shown in [**109**] that if a G-space R is a straight space (has unique metric segments) with convex spheres, then under the above assumptions, ϕ has a fixed point.

It was subsequently shown in [**112**] that it suffices to assume only that some subsequence of $\{\phi^n(x)\}$ is bounded in the preceding theorem, an assumption later shown by A. Całka [**48**] to be equivalent to the original. In fact he proves that in any finitely totally bounded metric space X and nonexpansive $f : X \to X$, boundedness of some subsequence of $\{f^n(x)\}$ for $x \in X$ implies boundedness of $\{f^n(x)\}$. This is a fact that is known to be false, for example, in a Hilbert space (see [**68**]).

A loop at a point p in G-space is a geodesic monogon L such that L is the union of two segments from p to q ($q \in L$) have only p and q in common. Let $\lambda(L)$ denote the length of a loop $L \subset G$, and denote by $Q(p)$ the set of all loops at $p \in R$. If $Q(p) \neq \emptyset$, set

$$\lambda_i(p) = \inf_{L \in Q(p)} \lambda(L); \quad \lambda_s(p) = \sup_{L \in Q(p)} \lambda(L)$$

and for $Q(p) = \emptyset$, set $\lambda_i(p) = \infty$; $\lambda_s(p) = 0$. Let

$$\lambda_i(R) = \inf_{p \in R} \lambda_i(p); \quad \lambda_s(R) = \sup_{p \in R} \lambda_s(p).$$

THEOREM 8.5 ([**111**]). *A G-space R does not possess proper local isometries if $\lambda_i(R) > 0$ and $\lambda_s(R) < \infty$.*

Całka's result has arisen again in several related contexts; for example see [**142**].

8.1. A Fundamental Problem in G-Spaces

Busemann proved that every one and two-dimensional G-space is a topological manifold, and he states [**46**, p. 49]:

> Although this is probably true for any G-space, the proof (if the conjecture is correct) seems quite inaccessible in the present state of topology.

We do not know the current state of this conjecture. However it was soon established for the case of three dimensions by B. Krakus [**134**], and in a surprising more recent development,[2] P. Thurston [**209**] established Busemann's conjecture for four dimensions.

THEOREM 8.6 (Berestovskii [**21**]). *Busemann G-spaces of dimension $n \geq$ 5 having Aleksandrov curvature bounded above*[3] *are n-manifolds.*

A Comment About Dimension.   There are various notions of dimension in topology (see, e.g., [**69**]). The one Busemann is referring to the classical "Menger-Urysohn" dimension. The axioms are:

(MU1) $\dim X = -1 \Leftrightarrow X = \emptyset$;

(MU2) $\dim X \leq n$, $n \in \mathbb{N}$, if for every point $x \in X$ and each neighborhood V_x of x there exists an open set $U \subset X$ such that

$$x \in U \subset V_x \text{ and } \dim FrU \leq n - 1;$$

(MU3) $\dim X = n$ if $\dim X \leq n$ and $\dim X > n - 1$, i.e., $\dim X$ is not $\leq n - 1$;

(MU4) $\dim X = \infty$ if $\dim X > n \; \forall n \in \mathbb{N}$.

In the realm of separable metric spaces (e.g., G-spaces) this concept of dimension coincides with the notion of "covering" dimension. See [**22**] for a recent survey of all known results on the topology of Busemann G-spaces of finite dimension.

---

[2]The reviewer of this paper states in [**188**]: "Without any doubt this is one of the nicest papers in geometric topology of the 1990's."

[3]A metric space is said to have Alexandrov curvature $\leq \kappa$ if it is locally a CAT(κ) space. CAT(κ) spaces are defined in the next chapter.

CHAPTER 9

CAT(0) Spaces

9.1. Introduction

A substantial part of the discussion in this chapter is taken from two survey articles [**118, 119**]. These articles motivated a substantial resurgence of the study of metric fixed point theory in spaces of non-positive curvature. The study of spaces of non-positive curvature originated with the discovery of hyperbolic spaces, the work of J. Hadamard at the beginning of the last century, and the work of E. Cartan in the 1920s. The idea of what it means for a geodesic metric space to have non-positive curvature (or, more generally, curvature bounded above by a real number κ) goes back to the work of H. Busemann and A.D. Alexandrov in the 1950s. Of particular importance to the revival of interest in this topic are the lectures which Mikhael Gromov gave in 1981 at the Collège de France in Paris (see [**36**, p. VIII]). In these lectures Gromov explained the main features of global Riemannian geometry essentially by basing his account wholly on the so-called CAT(0) inequality.

It is shown in [**118**] that many of the standard ideas and methods of nonlinear analysis and Banach space theory carry over to the class of spaces Gromov calls CAT(0) spaces. (The letters C, A, and T stand for Cartan, Alexandrov, and Toponogov.) A metric space X is said to be a *CAT(0) space* if it is geodesically connected, and if every geodesic triangle in X is at least as "thin" as its comparison triangle in the Euclidean plane. We make this precise below. However it is this latter property, known as the CAT(0) inequality, that encapsulates the concept of non-positive curvature in Riemannian geometry and allows one to reflect the same concept in a much wider setting. We shall almost invariably assume completeness as well. Complete CAT(0) spaces are often called *Hadamard spaces*. CAT(0) spaces have a remarkably nice geometric structure. One can see almost immediately that in such spaces angles exist in a strong sense, the distance function is convex, one has both uniform convexity and orthogonal projection onto convex subsets, etc. Also, because of their generality, CAT(0) spaces arise in a wide variety of contexts. In CAT(0) spaces the nonexpansive mappings arise naturally in the study of isometries or, more generally, local isometries.

© Springer International Publishing Switzerland 2014
W. Kirk, N. Shahzad, *Fixed Point Theory in Distance Spaces*,
DOI 10.1007/978-3-319-10927-5_9

In [186], Reich and Shafrir introduced a class of "hyperbolic" metric spaces which they proposed as "an appropriate background for the study of nonlinear operator theory in general, and of iterative processes for nonexpansive mappings in particular." The observations in this chapter should serve to reinforce that assessment. Within the hyperbolic framework, the CAT(0) spaces might be viewed as an analog to the Hilbert spaces in the classical theory of nonlinear analysis. However such an analogy could be misleading. CAT(0) spaces include all $\mathbb{R}$-trees, and these spaces bear little resemblance to Hilbert spaces.

A fundamental source for much of our exposition is the recent book [36] by M.R. Bridson and A. Haefliger, and one should look there for things not specifically attributed here to other sources. A more elementary treatment of many of these ideas can be found in the recent text of Burago et al. [45]. Many results, most of which are found in [36], are stated here without proof.

Several new results concerning fixed point theorems in CAT(0) spaces are also discussed and proofs of a few new results are included. We also indicate how some new fixed point theorems in $\mathbb{R}$-trees have applications to graph theory.

9.2. CAT(κ) Spaces

Denote by M_κ^n the following classical metric spaces:

(1) if $\kappa = 0$, then M_0^n is the Euclidean space $\mathbb{R}^n$;
(2) if $\kappa > 0$, then M_κ^n is obtained from the sphere $\mathbb{S}^n$ by multiplying the spherical distance by $1/\sqrt{\kappa}$;
(3) if $\kappa < 0$, then M_κ^n is obtained from the hyperbolic space $\mathbb{H}^n$ by multiplying the hyperbolic distance by $1/\sqrt{-\kappa}$.

A *geodesic triangle* $\Delta(x_1, x_2, x_3)$ in a geodesic metric space (X, d) consists of three points in X (the *vertices* of Δ) and a geodesic segment between each pair of vertices (the *edges* of Δ). A *comparison triangle* for geodesic triangle $\Delta(x_1, x_2, x_3)$ in (X, d) is a triangle $\overline{\Delta}(x_1, x_2, x_3) := \Delta(\bar{x}_1, \bar{x}_2, \bar{x}_3)$ in M_κ^2 such that $d_{\mathbb{R}^2}(\bar{x}_i, \bar{x}_j) = d(x_i, x_j)$ for $i, j \in \{1, 2, 3\}$. If $\kappa > 0$ it is further assumed that the perimeter of $\Delta(x_1, x_2, x_3)$ is less than $2D_\kappa$, where D_κ denotes the diameter of M_κ^2. Such a triangle always exists.

A geodesic metric space is said to be a CAT(κ) space if all geodesic triangles of appropriate size satisfy the following CAT(κ) comparison axiom.

CAT(κ): Let Δ be a geodesic triangle in X and let $\overline{\Delta} \subset M_\kappa^2$ be a comparison triangle for Δ. Then Δ is said to satisfy the CAT(κ) *inequality* if for all $x, y \in \Delta$ and all comparison points $\bar{x}, \bar{y} \in \overline{\Delta}$,

$$d(x, y) \leq d(\bar{x}, \bar{y}).$$

Complete CAT(0) spaces are often called *Hadamard spaces*. These spaces are of particular relevance to this study.

DEFINITION 9.1. A metric space X is said to be of *curvature* ≤ 0 if it is locally a CAT(0) space. In this case X is said to be *non-positively curved*.

The significance of the above definition lies in the fact that it provides a good notion of an upper bound on curvature in an arbitrary geodesic space. In fact, classical theorems in differential geometry show that if a Riemannian manifold is sufficiently smooth (e.g., C^3), then it has curvature in the above sense if and only if its sectional curvatures are $\leq \kappa$. This is a result due to Alexandrov [**8**] in general, and to E. Cartan [**51**] in the case $\kappa = 0$. It should be noted that nonpositively curved spaces play a significant role in many branches of mathematics. See, e.g., B. Kleiner's review [**132**] of [**36**].

To continue with the terminology of [**36**], the metric on a space X is said to be *convex* if X is a geodesic space and all geodesics $c_1 : [0, a_1] \to X$ and $c_2 : [0, a_2] \to X$ with $c_1(0) = c_2(0)$ satisfy the inequality

$$d(c_1(ta_1), c_2(ta_2)) \leq td(c_1(a_1), c_2(a_2))$$

for all $t \in [0, 1]$. X is said to be *locally convex* if every point has a neighborhood in which the induced metric is convex. If the metric space is locally convex, then in particular X is locally contractible, and therefore X has a universal covering space $\tilde{X}$. This means that X is simply connected and there is a local isometry $p : \tilde{X} \to X$. In fact the space $\tilde{X}$ is unique up to an isometry.

THEOREM 9.1 ((Cartan–Hadamard Theorem) [**36**, p. 193]). *Let X be a complete and connected metric space.*

(1) *If the metric on X is locally convex, then the induced length metric on the universal covering space $\tilde{X}$ is (globally) convex.*

(2) *If X is of curvature ≤ 0, then $\tilde{X}$ is of CAT(0).*

Thus if a complete simply connected length space is locally convex and has curvature ≤ 0, it is a CAT(0) space.

The CAT(0) inequality may be stated in the following equivalent but formally weaker form.

PROPOSITION 9.1 (cf. [**36**, p. 161]). *A geodesic metric space is a CAT(0) space if and only if the following condition holds: For every geodesic triangle $\Delta(p, q, r)$ in X and every point $x \in [q, r]$ the following inequality is satisfied by the comparison point $\bar{x} \in [\bar{q}, \bar{r}] \subset \overline{\Delta}(p, q, r) \subset \mathbb{R}^2$:*

$$d(p, x) \leq d(\bar{p}, \bar{x}).$$

We now collect some further properties of CAT(κ) spaces. These are all found in [**36**].

PROPOSITION 9.2. *Let X be a CAT(κ) space.*

(1) *There is a unique geodesic segment joining each pair of points $x, y \in X$ (provided $d(x, y) < D_\kappa$ if $\kappa > 0$), and this geodesic segment varies continuously with its endpoints.*

(2) *Every local geodesic in X of length at most D_κ is a geodesic.*
(3) *The balls in X of radius smaller that $D_\kappa/2$ are convex.*
(4) *The balls in X of radius less than D_κ are contractible.*
(5) *Approximate midpoints are close to midpoints. Specifically, for every $\lambda < D_\kappa$ and $\varepsilon > 0$ there exists $\delta = \delta(\kappa, \lambda, \varepsilon)$ such that if m is the midpoint of a geodesic segment $[x, y] \subset X$ with $d(x, y) \leq \lambda$ and if*

$$\max\{d(x, m'), d(y, m')\} \leq \frac{1}{2}d(x, y) + \delta,$$

then $d(m, m') < \varepsilon$.

THEOREM 9.2. *The following relations hold:*

(1) *If X is a CAT(κ) space, then it is a CAT(κ') space for every $\kappa' \geq \kappa$.*
(2) *If X is a CAT(κ') space for every $\kappa' > \kappa$, then it is a CAT(κ) space.*
(3) *X is a CAT(κ) space for all κ if and only if X is an $\mathbb{R}$-tree.*

One consequence of (1) is that any result proved for CAT(0) spaces automatically carries over to CAT(κ) spaces for $\kappa < 0$, and especially to $\mathbb{R}$-trees. (See Chap. 11 for the definition of an $\mathbb{R}$-tree.)

The following is another important observation.

PROPOSITION 9.3 ([**36**, p. 176]). *If X is a CAT(0) space, then the distance function $d : X \times X \to \mathbb{R}$ is convex.*

This means that given any pair of geodesics $c : [0, 1] \to X$ and $c' [0, 1] \to X$ parametrized proportional to arc length, the following inequality holds for all $t \in [0, 1]$:

$$d(c(t), c'(t)) \leq (1 - t) d(c(0), c'(0)) + td(c(1), c'(1)).$$

We now turn to a generalization of Jung's theorem. (This result also holds for spaces of curvature bounded below.) We use $rad(S)$ to denote the usual Chebyshev radius of S relative to the underlying space X:

$$rad(S) = \inf\{\rho > 0 : S \subseteq B(x; \rho) \text{ for some } x \in X\}.$$

We also need to introduce the function $sn_\kappa : \mathbb{R} \to \mathbb{R}$, defined by

$$sn_\kappa(x) := \begin{cases} \sin(\sqrt{\kappa}x)/\sqrt{\kappa} & \text{if } \kappa > 0, \\ x & \text{if } \kappa = 0, \\ \sinh(\sqrt{-\kappa}x)/\sqrt{-\kappa} & \text{if } \kappa < 0. \end{cases}$$

THEOREM 9.3 ([**137**]). *Let X be a complete CAT(κ) space and S a nonempty bounded subset of X. Then there exists a unique $x \in X$ such that $S \subseteq B(x; rad(S))$, where*

$$sn_\kappa(rad(S)) \leq \sqrt{2}sn_\kappa\left(\frac{diam(S)}{2}\right).$$

In particular (cf., [47]), if S is a bounded subset of a complete CAT(0) space, then

$$rad\left(S\right) \le \frac{\sqrt{2}}{2} diam\left(S\right).$$

This of course coincides with the well-known Hilbert space estimate.

While many reformulations of the CAT(κ) condition concern the geometry of triangles, there is also a useful reformulation involving the geometry of quadrilaterals. Let (x_1, x_2, x_3, x_4) be a 4-tuple of point of a metric space X. A *subembedding* of this 4-tuple in M_κ^2 is a 4-tuple $(\bar{x}_1, \bar{y}_1, \bar{x}_2, \bar{y}_2)$ of points of M_κ^2 such that $d\left(\bar{x}_i, \bar{y}_j\right) = d\left(x_i, y_j\right)$ for $i, j \in \{1, 2\}$, and

$$d\left(x_1, x_2\right) \le d\left(\bar{x}_1, \bar{x}_2\right) \text{ and } d\left(y_1, y_2\right) \le d\left(\bar{y}_1, \bar{y}_2\right).$$

X is said to satisfy the CAT(κ) 4-point condition if every 4-tuple of points

$$\left(x_1, y_1, x_2, y_2\right)$$

in X for which

$$d\left(x_1, y_1\right) + d\left(y_1, x_2\right) + d\left(x_2, y_2\right) + d\left(y_2, x_1\right) < 2D_\kappa$$

has a subembedding in M_κ^2.

PROPOSITION 9.4. *Let X be a complete metric space. Then the following conditions are equivalent.*

(1) X *is a CAT(κ) space.*
(2) X *satisfies the CAT(κ) 4-point condition and every pair of points $x, y \in X$ with $d\left(x, y\right) < D_\kappa$ has approximate midpoints.*

Finally we observe that if x, y_1, y_2 are points of a CAT(0) space and if y_0 is the midpoint of the segment $[y_1, y_2]$, then the CAT(0) inequality implies

$$(9.1) \qquad d\left(x, y_0\right)^2 \le \frac{1}{2} d\left(x, y_1\right)^2 + \frac{1}{2} d\left(x, y_2\right)^2 - \frac{1}{4} d\left(y_1, y_2\right)^2$$

because equality holds in the Euclidean metric. In fact (cf., [36, p. 163]), *a geodesic metric space is a CAT(0) space if and only if it satisfies inequality (9.1)* (which is known as the CN *inequality* of Bruhat and Tits [41]). Using this inequality it is easy to see that if X_1 and X_2 are CAT(0) spaces, then $X_1 \times X_2$ is also a CAT(0) space, where the metric on $X_1 \times X_2$ is given by

$$d\left(\left(x_1, x_2\right), \left(y_1, y_2\right)\right)^2 = d\left(x_1, y_1\right)^2 + d\left(x_2, y_2\right)^2.$$

Also if $d\left(x, y_1\right) = d\left(x, y_2\right) = 1$ and $d\left(y_1, y_2\right) \ge \varepsilon$, then (9.1) gives

$$d\left(x, y_0\right) \le \sqrt{1 - \frac{\varepsilon^2}{4}}$$

and so one has the usual Euclidean modulus of convexity in CAT(0) spaces. In particular, $d(x, y_1) \leq R$, $d(x, y_2) \leq R$, and $d(y_1, y_2) \geq r$ imply

$$d(x, y_0) \leq \left(1 - \delta\left(\frac{r}{R}\right)\right) R,$$

where $\delta(\varepsilon) := \sqrt{1 - \dfrac{\varepsilon^2}{4}}$.

An extremely useful property of CAT(0) is the nearest point projection. Crucial to the proof of this fact, and useful in other contexts as well, is the concept of "angle" in a metric space. Let X be a metric space, and let $c : [0, a] \to X$ and $c' : [0, a'] \to X$ be two geodesics with $c(0) = c'(0)$. Given $t \in (0, a]$ and $t' \in (0, a']$ we consider the comparison triangle $\bar{\Delta}(c(0), c(t), c'(t'))$ and the comparison angle $\bar{\angle}_{c(0)}(c(t), c'(t'))$. The (Alexandrov) *angle* between the geodesic paths c and c' is the number $\angle_{c,c'} \in [0, \pi]$ defined by:

$$\angle_{c,c'} = \limsup_{t,t' \to 0} \bar{\angle}_{c(0)}(c(t), c'(t')) = \lim_{\varepsilon \to 0} \sup_{0 < t, t' < \varepsilon} \bar{\angle}_{c(0)}(c(t), c'(t')).$$

PROPOSITION 9.5 ([36, p. 176]). *Let X be a CAT(0) space, and let C be a convex subset of X which is complete in the induced metric. Then:*

(1) *for every $x \in X$ there exists a unique point $P(x) \in C$ such that*

$$d(x, P(x)) = dist(x, C);$$

(2) *if x' belongs to the geodesic segment $[x, P(x)]$, then $P(x') = P(x)$;*

(3) *given $x \notin C$ and $y \in C$, if $y \neq P(x)$ then $\angle_{P(x)}(x, y) \geq \pi/2$;*

(4) *the map $x \mapsto P(x)$ is a nonexpansive retraction of X onto C; the map $H : X \times [0, 1] \to X$ associating with (x, t) the point at distance $td(x, P(x))$ from x on the geodesic segment $[x, P(x)]$ is a continuous homotopy from the identity map of X to P.*

9.3. Fixed Point Theory

We now come to one of the central topics of this monograph. From the preceding section it is easy to see that CAT(0) spaces share many properties of uniformly convex Banach spaces. For example, closed convex sets are uniquely proximinal, descending sequences of nonempty bounded closed convex sets have nonempty intersection, and "asymptotic center" techniques apply. As we have seen, CAT(0) spaces also enjoy certain Hilbert space properties: For example, nearest point projections onto closed convex sets are nonexpansive, and a notion of angle is present for which a law of cosines applies. Also the family of all bounded closed convex subsets of a given CAT(0) space is normal in the sense described in Chap. 3. Thus the following theorem is immediate.

THEOREM 9.4. *Suppose K is a nonempty bounded closed convex subset of a complete CAT(0) space and suppose $f : K \to K$ is nonexpansive. Then the set of fixed points of f is nonempty, closed, and convex.*

We begin with some general notation. Let S be a subset of a complete CAT(0) space X. Then for each $x, y \in S$ there is a unique geodesic $[x, y] \subset X$ joining x and y. We denote by $G_1(S)$ the union of all geodesic segments in X with endpoints in S. Then S is *convex* if $G_1(S) = S$. The CAT(0) inequality ensures that $rad(G_1(S)) = rad(S)$. For $n \geq 2$, define inductively $G_n(S) = G_1(G_{n-1}(S))$. The *convex hull* of S is defined to be the set

$$conv S = \bigcup_{n=1}^{\infty} G_n(S),$$

$\overline{conv}S$ denotes its closure. From this we conclude that $rad(\overline{conv}S) = rad(S)$ for every bounded $S \subset X$.

One of the fundamental theorems in fixed point of nonexpansive mappings is the demiclosedness theorem due to F. Browder [**40**]. This theorem asserts that if K is a closed and convex subset of a uniformly convex Banach space X, and if $f : K \to X$ is nonexpansive, then $I - f$ is *demiclosed* on K, that is, if $\{u_j\}$ is a sequence in K which converges weakly to $u \in X$ and if $\{(I - f)(u_j)\}$ converges strongly (in norm) to w, then $w \in K$ and $(I - f)(u) = w$. One important corollary of this theorem is that if $\inf\{\|x - f(x)\| : x \in K\} = 0$, then f has a fixed point in K when K is bounded. In the absence of a weak topology, an analog of this theorem cannot even be formulated in a complete CAT(0) space. However the corollary can be formulated in such a setting, and indeed, it turns out to be true.

We need some notation. Given a mapping $f : K \to X$ where K is a subset of a metric space X, and a number $\varepsilon > 0$, the *ε-fixed point* set of f is the set

$$F_\varepsilon(f) = \{x \in K : d(x, f(x)) \leq \varepsilon\}.$$

We take the following lemma as a point of departure.

LEMMA 9.1 ([**36**, p. 286]). *Let X be a CAT(0) space. Fix $x, y \in X$ with $d(x, y) = 2r$ and let m be the midpoint of the geodesic segment $[x, y]$. If m' is a point such that $d(m', x) \leq r(1 + \varepsilon)$ and $d(m', y) \leq r(1 + \varepsilon)$, then $d(m, m') \leq r\sqrt{\varepsilon^2 + 2\varepsilon}$.*

A slight modification of the argument given in [**36**] for Lemma 9.1 leads to the following result.

LEMMA 9.2. *Let K be a bounded subset of a CAT(0) space X, suppose $f : K \to X$ is nonexpansive, and suppose $x, y \in F_\varepsilon(f)$ with $d(x, y) = r$. Let $m \in [x, y] \cap K$. Then $f(m) \in F_{\phi(\varepsilon)}(f)$, where $\phi(\varepsilon) = \sqrt{\varepsilon^2 + 2r\varepsilon}$.*

PROOF. Let m be the point of $[x, y]$ with distance αr from x, and suppose $m \in K$. Then if $m' = f(m)$,

$$\begin{aligned} d(x, m') &\leq d(x, f(x)) + d(f(x), f(m)) \\ &\leq d(x, f(x)) + d(x, m) \\ &\leq \varepsilon + \alpha r. \end{aligned}$$

Similarly $d\left(y, m'\right) \leq \varepsilon + (1 - \alpha)\, r$. At least one of the angles $\angle_m\left(m', x\right)$ and $\angle_m\left(m', y\right)$ is greater than or equal to $\pi/2$. If $\angle_m\left(m', x\right) \geq \pi/2$, then in the comparison triangle $\overline{\Delta}\left(m, m', x\right)$ the angle at $\bar{m}$ is also greater than or equal to $\pi/2$. By the law of cosines

$$\left(\varepsilon + \alpha r\right)^2 \geq d\left(x, m'\right)^2 \geq \left(\alpha r\right)^2 + d\left(m, m'\right)^2 .$$

Similarly, if $\angle_m\left(m', y\right) \geq \pi/2$,

$$\left(\varepsilon + (1 - \alpha)\, r\right)^2 \geq d\left(y, m'\right)^2 \geq \left((1 - \alpha)\, r\right)^2 + d\left(m, m'\right)^2 .$$

Therefore

$$d\left(m, f\left(m\right)\right)^2 = d\left(m, m'\right)^2 \leq \max\left\{\varepsilon^2 + 2\alpha r\varepsilon, \varepsilon^2 + 2\left(1 - \alpha\right) r\varepsilon\right\}$$
$$\leq \varepsilon^2 + 2r\varepsilon.$$

$\square$

THEOREM 9.5. *Let K be a bounded closed convex subset of a complete CAT(0) space X. Suppose $f : K \to X$ is a nonexpansive mapping for which*

$$\inf\left\{d\left(x, f\left(x\right)\right) : x \in K\right\} = 0.$$

Then f has a fixed point in K.

PROOF. Let $x_0 \in X$ be fixed and define

$$r_0 = \inf\left\{r > 0 : \inf\left\{d\left(x, f\left(x\right)\right) : x \in B\left(x_0; r\right) \cap K\right\} = 0\right\}.$$

Obviously $r_0 < \infty$, and if $r_0 = 0$, then $x_0 \in K$ and $f\left(x_0\right) = x_0$ by continuity of f. So we suppose $r_0 > 0$. Now choose $\left\{x_n\right\} \subset K$ so that $d\left(x_n, f\left(x_n\right)\right) \to 0$ and $d\left(x_0, x_n\right) \to r_0$. Since any convergent subsequence of $\left\{x_n\right\}$ would have a fixed point of f as its limit, we may suppose there exist $\varepsilon > 0$ and subsequences $\left\{u_j\right\}$ and $\left\{v_j\right\}$ of $\left\{x_n\right\}$ such that $d\left(u_j, v_j\right) \geq \varepsilon$. Passing again to subsequences if necessary we may also suppose $d\left(x_0, u_j\right) \leq r_0 + \dfrac{1}{j}$ and $d\left(x_0, v_j\right) \leq r_0 + \dfrac{1}{j}$. Let m_j be the midpoint of the segment $[u_j, v_j]$ and let $\bar{m}_j$ be the point corresponding to m_j on the comparison triangle $\Delta\left(\bar{x}_0, \bar{u}_j, \bar{v}_j\right)$. Then by the CAT(0) inequality

$$d\left(x_0, m_j\right) \leq d\left(\bar{x}_0, \bar{m}_j\right) \leq \sqrt{\left(r_0 + \dfrac{1}{j}\right)^2 - \left(\dfrac{\varepsilon}{2}\right)^2}.$$

Clearly $d\left(x_0, m_j\right) \leq r^* < r_0$ for j sufficiently large. On the other hand, by Lemma 9.2, $d\left(m_j, f\left(m_j\right)\right) \to 0$ as $j \to \infty$. This contradicts the definition of r_0. $\square$

Essentially the same proof gives the following result.

THEOREM 9.6. *Let U be a connected bounded open set in a complete CAT(0) space X, and let $f : \overline{U} \to X$ be nonexpansive. Then the following alternative holds:*

(a) f has a fixed point in U, or;

(b) $\inf\{d(x, f(x)) : x \in \partial U\} \leq \inf\{d(x, f(x)) : x \in U\}$.

PROOF. Assume there exists $z \in U$ such that

$$d(z, f(z)) < \xi := \inf\{d(x, f(x)) : x \in \partial U\}.$$

Then if $m \in [z, f(z)] \cap U$,

$$\begin{aligned} d(m, f(m)) &\leq d(m, f(z)) + d(f(z), f(m)) \\ &\leq d(m, f(z)) + d(z, m) \\ &= d(z, f(z)). \end{aligned}$$

This proves that the segment $[z, f(z)]$ not only lies in U but in fact it is bounded away from ∂U. Consequently if one defines $g : U \to X$ by taking $g(x)$ to be the midpoint of the segment $[x, f(x)]$ for each $x \in U$, the sequence $\{g^n(z)\}$ lies in U. Moreover by Theorem 6.1, $d(g^n(z), g^{n+1}(z)) \to 0$. Thus $\inf\{d(x, f(x)) : x \in U\} = 0$. The argument now follows the preceding one. All one needs to observe is that if $\varepsilon > 0$ is chosen so that $\sqrt{\varepsilon^2 + 2r\varepsilon} \leq \delta < \xi$, where $r = diam(U)$, and if $m \in [u, v] \cap U$, where $u, v \in F_\varepsilon(f)$, then $d(m, f(m)) \leq \delta$. Hence $[u, v] \cap \partial U = \emptyset$. Thus the points m_k as defined in the preceding argument all lie in U. $\qquad\square$

REMARK 9.1. *It is shown in* [119] *that Theorem 9.6 holds under the weaker assumption that* $f : \overline{U} \to X$ *is continuous on* $\overline{U}$ *and locally nonexpansive on* D.

We next consider the approximate fixed point property. A subset K of a metric space is said to have the *approximate fixed point property* (for nonexpansive mappings) if given any nonexpansive $f : K \to K$, $\inf\{d(x, f(x)) : x \in K\} = 0$. To characterize this concept, we need some more definitions.

DEFINITION 9.2 ([197]). Let X be a metric space. A curve $\gamma : [0, \infty) \to X$ is said to be *directional* (with constant b) if there is $b \geq 0$ such that

$$t - s - b \leq d(\gamma(s), \gamma(t)) \leq t - s$$

for all $t \geq s \geq 0$. A subset of X is said to be *directionally bounded* if it does not contain a directional curve.

DEFINITION 9.3. A geodesic metric space X is said to have the *geodesic extension property* if for every local geodesic $c : [a, b] \to X$, with $a \neq b$, there exists $\varepsilon > 0$ and a local geodesic $c' : [a, b + \varepsilon] \to X$ such that $c'|_{[a,b]} = c$.

LEMMA 9.3 ([36, p. 298]). *If* X *is a* $CAT(0)$ *space, then* X *has the geodesic extension property if and only if every non-constant geodesic* $c : [a, b] \to X$ *can be extended to a line* $c : \mathbb{R} \to X$.

(If a complete CAT(0) space is homeomorphic to a finite dimensional manifold, then it always has the geodesic extension property.)

In [185] it is shown that a reflexive Banach space has the approximate fixed point property if and only if it is directionally bounded (i.e., its intersection with any line is bounded), and in [197] Shafrir proved that a closed convex subset of a complete hyperbolic metric space has the approximate fixed point property for nonexpansive mappings if and only if it is directionally bounded. As an immediate corollary of Shafrir's result, a closed convex subset of a complete CAT(0) space with the geodesic extension property has the approximate fixed point property if and only if it is directionally bounded. However in this case the stronger assertion of Reich's result is true. (One should also compare this result to Theorem 32.2 of [88], which states that the same result holds for the complex Hilbert ball with a hyperbolic metric. In fact, in this setting the approximate fixed point property actually implies the fixed point property.)

THEOREM 9.7. *A closed convex subset of a complete CAT(0) space with the geodesic extension property has the approximate fixed point property for nonexpansive mappings if and only if it does not contain a geodesic ray.*

PROOF. In view of Shafrir's result it need only be shown that if a closed convex set in a complete CAT(0) space X is geodesically bounded then it is directionally bounded. So, suppose K is a closed convex set in X and suppose K contains a directional curve γ. We show that this implies K contains a geodesic ray.

Let $x_n = \gamma(n)$, $n = 0, 1, 2, \cdots$, and fix an arbitrary $\rho > b$, where b is the directional constant associated with γ. For each $n \geq \rho$, let y_n be the point of geodesic segment $[x_0, x_n]$ with distance ρ from x_0. Now suppose $m > n \geq \rho$, and let $\alpha_{n,m}$ be the comparison angle $\angle_{\bar{x}_0}(\bar{x}_n, \bar{x}_m)$ in $\mathbb{R}^2$. By the law of cosines

$$\cos(\alpha_{n,m}) = \frac{d(x_0, x_n)^2 + d(x_0, x_m)^2 - d(x_n, x_m)^2}{2d(x_0, x_n)d(x_0, x_m)}.$$

Using the inequalities

$$n - b \leq d(x_0, x_n) \leq n;$$

$$m - b \leq d(x_0, x_m) \leq m;$$

$$m - n \geq d(x_n, x_m),$$

we have

$$\cos{(\alpha_{n,m})} \geq \frac{(n-b)^2 + (m-b)^2 - (m-n)^2}{2nm}$$

$$= \frac{n}{2m}\left[\frac{(n-b)^2}{n^2}\right] + \frac{m}{2n}\left[\frac{(m-b)^2}{m^2}\right] - \frac{(m-n)^2}{2nm}$$

$$= \frac{n}{2m} + \frac{m}{2n} - \frac{(m-n)^2}{2nm} - \frac{b}{n} - \frac{b}{m} + \frac{b^2}{nm}$$

$$= 1 - b\left(\frac{1}{n} + \frac{1}{m} - \frac{b}{nm}\right).$$

Thus $\cos{(\alpha_{n,m})} \to 1$ as $m, n \to \infty$; hence $\alpha_{n,m} \to 0$. If $\bar{y}_n, \bar{y}_m$ are the respective points of the comparison triangle $\Delta\left(\bar{x}_0, \bar{x}_n, \bar{x}_m\right)$ corresponding to y_n, y_m, then by the CAT(0) inequality $d\left(y_n, y_m\right) \leq d\left(\bar{y}_n, \bar{y}_m\right)$. The fact that $\alpha_{n,m} \to 0$ as $m, n \to \infty$ implies that $\{\bar{y}_n\}$, hence $\{y_n\}$, is a Cauchy sequence. Since $\rho > b$ is arbitrary it now follows that the sequence $\{[x_0, x_n]\}$ of geodesic segments converges to a geodesic ray issuing from x_0. $\qquad\square$

It has been known for some time that a nonempty closed convex subset of a Hilbert space has the fixed point property for nonexpansive mappings if and only if it is bounded (Ray's theorem [**183**]). In view of this it might be tempting to conjecture that a closed convex subset of a complete CAT(0) space has the fixed point property if and only if it is bounded. However this is false. In [**73**] it is shown that a closed convex subset of an $\mathbb{R}$-tree has the fixed point property for nonexpansive mappings if (and only if) it is *geodesically bounded*. The question of whether there is a class of CAT(0) spaces for which Ray's theorem holds is taken up in [**76**]. An affirmative answer is given via the introduction of the following concept:

DEFINITION 9.4. Let (X, d) be an unbounded geodesic space. Then X is said to have the *property of the far unbounded set* (*property U for short*) if for any closed convex unbounded set $Y \subset X$, either Y is geodesically unbounded, or for each closed convex unbounded set $K \subseteq Y$ and $x \in K$ there exists a closed convex unbounded subset K_1 of K such that

$$dist\left(x, K_1\right) := \inf\left\{d\left(x, y\right) : y \in K_1\right\} \geq 1.$$

One of the central results of [**76**] asserts that if (X, d) is a complete CAT(0) space possessing property U, then a closed convex subset Y of X has the fixed point property for nonexpansive mappings if and only if it is bounded. (It is also shown in [**76**] that any reflexive Banach space has property U.)

We now turn to a method for approximating fixed points of nonexpansive mappings in CAT(0) spaces. Let K be a bounded closed convex subset of a complete CAT(0) space and suppose $f : K \to K$ is nonexpansive. Fix $x_0 \in K$, and define the mapping $f_t : K \to K$ by taking $f_t(u)$, $t \in (0,1)$, to be the point of $[x_0, f(u)]$ at distance $td(x_0, f(u))$ from x_0. Then by convexity of the metric

$$d(f_t(u), f_t(v)) \leq td(u,v),$$

so f_t is a contraction mapping of K into K. Since K is complete, Banach's contraction mapping theorem assures the existence of a unique point x_t such that:

(9.2) $\qquad\qquad x_t \in [x_0, f(x_t)] \text{ and } d(x_0, x_t) = td(x_0, f(x_t)).$

This fact can be used to prove the following theorem. For an analog of this result in the Hilbert ball, see Theorem 24.1 of [88].

THEOREM 9.8. *Let K be a bounded closed convex subset of a complete CAT(0) space X, let $f : K \to K$ be nonexpansive, fix $x_0 \in K$, and for each $t \in [0,1)$ let x_t be the point of $[x_0, f(x_t)]$ satisfying (9.2). Then $\lim_{t \to 1^-} x_t$ converges to the unique fixed point of f which is nearest x.*

PROOF. Fix $0 < j < l \leq 1$ and consider $\Delta\left(\bar{x}_0, \bar{f}(x_j), \bar{f}(x_l)\right)$, the comparison triangle of $\Delta(x_0, f(x_j), f(x_l))$ in $\mathbb{R}^2$. For convenience we take $\bar{x}_0$ to be the origin. By the CAT(0) inequality we have

$$\left\| \bar{f}(x_l) - \bar{f}(x_j) \right\| = d(f(x_l), f(x_j)) \leq d(x_l, x_j)$$
$$\leq \left\| \bar{x}_l - \bar{x}_j \right\| = \left\| j^{-1} \bar{f}(x_l) - l^{-1} \bar{f}(x_j) \right\|.$$

It is now possible to follow the argument of Halpern [91] step-by-step. Specifically, we have $\bar{x}_j = j\bar{f}(x_j)$ and $\bar{x}_l = l\bar{f}(x_l)$. Let $d = \bar{x}_l - \bar{x}_j$. Then

$$
\begin{aligned}
\left\langle l^{-1}(\bar{x}_j + d) - j^{-1}\bar{x}_j, l^{-1}(\bar{x}_j + d) - j^{-1}\bar{x}_j \right\rangle &= \left\| l^{-1}(\bar{x}_j + d) - j^{-1}\bar{x}_j \right\|^2 \\
&= \left\| l^{-1}\bar{x}_l - j^{-1}\bar{x}_j \right\|^2 \\
&= \left\| \bar{f}(x_l) - \bar{f}(x_j) \right\| \\
&\leq \|d\|^2.
\end{aligned}
$$

Thus

$$\left(l^{-1} - j^{-1}\right)^2 \|\bar{x}_j\|^2 + \left(l^{-1}\right)^2 \|d\|^2 + 2\langle\left(l^{-1} - j^{-1}\right)l^{-1}\bar{x}_j, d\rangle \leq \|d\|^2$$

from which

$$\left(l^{-1} - j^{-1}\right)^2 \|\bar{x}_j\|^2 + \left(l^{-2} - 1\right)\|d\|^2 \leq 2\left(j^{-1} - l^{-1}\right)l^{-1}\langle\bar{x}_j, d\rangle.$$

In particular $\langle\bar{x}_j, d\rangle \geq 0$. Now observe that

$$\|\bar{x}_l\|^2 = \langle\bar{x}_j + d, \bar{x}_j + d\rangle = \|\bar{x}_j\|^2 + \|d\|^2 + 2\langle\bar{x}_j, d\rangle.$$

Therefore

(9.3) $\qquad\qquad \|\bar{x}_l\|^2 \geq \|\bar{x}_j\|^2 + \|\bar{x}_j - \bar{x}_l\|^2.$

Now let $\{k_i\}$ satisfy $0 < k_1 < k_2 < \cdots < 1$ with $k_i \to 1$ as $i \to \infty$. Since the sequence $\left\{ \left\| \bar{x}_{k_i} \right\|^2 \right\}$ is monotone increasing and bounded, it follows that

$$d\left(x_{k_i}, x_{k_j}\right)^2 \leq \left\| \bar{x}_{k_i} - \bar{x}_{k_j} \right\|^2 \leq \left\| \bar{x}_{k_i} \right\|^2 - \left\| \bar{x}_{k_j} \right\|^2 \to 0 \text{ as } i, j \to \infty.$$

Thus $\{x_{k_i}\}$ converges to some point $x^* \in K$. However

$$d\left(x_{k_i}, f\left(x_{k_i}\right)\right) = (1 - k_i)\, d\left(x_0, f\left(x_{k_i}\right)\right)$$

and by continuity

$$d\left(x^*, f\left(x^*\right)\right) = \lim_{k_i \to 1^-} (1 - k_i)\, d\left(x_0, f\left(x_{k_i}\right)\right) = 0.$$

Therefore x^* is a fixed point of f.

Now let p be any other fixed point of f. By repeating the preceding argument taking $x_l = x_1 = p$, we conclude from (9.3) that

$$\begin{aligned}
d\left(x_0, p\right)^2 &= \|\bar{p}\|^2 \geq \left\| \bar{x}_{k_i} \right\|^2 + \left\| \bar{x}_{k_i} - \bar{p} \right\|^2 \\
&\geq \left\| \bar{x}_{k_i} \right\|^2 + d\left(x_{k_i}, p\right)^2 \\
&= k_i^2 d\left(x_0, x_{k_i}\right)^2 + d\left(x_{k_i}, p\right)^2,
\end{aligned}$$

from which (letting $k_i \to 1^-$)

$$d\left(x_0, p\right)^2 \geq d\left(x_0, x^*\right)^2 + d\left(x^*, p\right)^2.$$

This proves that x^* is the unique fixed point of f which is nearest x_0. $\qquad\square$

As a consequence of the above result, if $fix\,(f)$ denotes the fixed point set of f, then given any $x \in K$,

$$\lim_{t \to 1^-} x_t = Px \in fix\,(f),$$

where the mapping P defined by $x \mapsto Px$ is the nearest point projection of K onto $fix\,(f)$. Since the nearest point projection of CAT(0) space X onto any complete convex subset of X is nonexpansive, P is nonexpansive. Thus $fix\,(f)$ is a nonexpansive retract of X.

In a Banach space context, the fact that in Theorem 9.8 is true outside Hilbert space, and indeed in any uniformly smooth space (except that the limit is a certain retraction different from the nearest point projection), is proved in Reich [184].

REMARK 9.2. *In fact, it is possible to show that the mapping P is firmly nonexpansive in the sense of Theorem 27.2 of [88].*

We now give two homotopy invariance results in the spirit of Frigon [84]. In these results $int\,K$ and ∂K denote, respectively, the interior and boundary of K.

THEOREM 9.9. *Let K be a closed convex subset of a complete $CAT(0)$ space X with $intK \neq \emptyset$. Suppose $f, g : K \to X$ are contraction mappings, and suppose $h : [0,1] \times K \to X$ is a homotopy satisfying*

    (a) *$h(0, \cdot) = g$; $h(1, \cdot) = f(\cdot)$;*
    (b) *$h(t, \cdot)$ is a contraction mapping with constant $k \in (0,1)$ for each $t \in [0,1]$;*
    (c) *For any $\varepsilon > 0$ there exists $\delta > 0$ such that for $x \in K$ and $t, s \in [0,1]$, $|t - s| < \delta \Rightarrow d(h(t, x), h(s, x)) < \varepsilon$;*
    (d) *$x \neq h(t, x)$ for each $x \in \partial K$ and $t \in [0,1]$.*

Then f has a fixed point in K if and only if g has a fixed point in K.

PROOF. Assume that G has a fixed point, and let

$$E = \{t \in [0,1] : x = h(t, x) \text{ for some } x \in K\}.$$

The theorem is proved by showing that E is both open and closed. If E is not open, then there exists $t \in E$ and $\{t_n\} \subset [0,1] \setminus E$ such that $t_n \to t$. Now let P be the nearest point projection of X onto K. Define the mappings $\hat{h}(t, \cdot) : X \to X$ by setting $\hat{h}(t, x) = h(t, P(x))$. Then the mappings $\hat{h}(t, \cdot)$ are contraction mappings with constant k. In particular there exist points $x_n \in X$ such that $x_n = \hat{h}(t_n, x_n)$. Thus

$$
\begin{aligned}
d(x_n, x) &= d(h(t_n, P(x_n)), h(t, P(x))) \\
&\leq d(h(t_n, P(x_n)), h(t_n, P(x))) + d(h(t_n, P(x)), h(t, P(x))) \\
&\leq kd(x_n, x) + d(h(t_n, P(x)), h(t, P(x))),
\end{aligned}
$$

from which

$$d(x_n, x) \leq (1 - k)^{-1} d(h(t_n, P(x)), h(t, P(x))).$$

By (c) it must be the case that $x_n = h(t_n, P(x_n)) \to x$. However $\{x_n\}$ is in the complement of K and x is in the interior of K. This contradiction proves that E is open.

To prove that E is closed, assume $\{t_n\} \subset E$ with $t_n \to t \notin E$. The same argument as the one just given leads to the conclusion that $x_n \to x$ with $\{x_n\}$ in the interior of K and x in the complement of K. ☐

THEOREM 9.10. *Let K be a bounded closed convex subset of a complete $CAT(0)$ space X with $intK \neq \emptyset$. Suppose $f : K \to X$ is a nonexpansive mapping, and suppose $h : [0,1] \times K \to X$ is a homotopy satisfying*

    (a) *$h(0, \cdot)$ has a fixed point;*
    (b) *$h(1, \cdot) = f(\cdot)$;*
    (c) *For each $t \in [0,1)$, $h(s, \cdot)$ is a k_t-contraction mapping for each $s \in [0, t)$;*
    (d) *For any $\varepsilon > 0$ there exists $\delta > 0$ such that for $x \in K$ and $t, s \in [0,1]$, $|t - s| < \delta \Rightarrow d(h(t, x), h(s, x)) < \varepsilon$;*
    (e) *$x \neq h(t, x)$ for each $x \in \partial K$ and $t \in [0,1]$.*

Then f has a fixed point in K.

PROOF. Select $\{t_n\} \subset (0,1)$ with $t_n \to 1$. By Theorem 9.9, for each n there exists $x_n \in K$ such that $x_n = h\left(t_n, x_n\right)$. Thus

$$d\left(x_n, f\left(x_n\right)\right) = d\left(h\left(t_n, x_n\right), h\left(1, x_n\right)\right),$$

and by (c) we conclude $d\left(x_n, f\left(x_n\right)\right) \to 0$. The result now follows from Theorem 9.5. $\qquad\square$

REMARK 9.3. *It is possible to prove that Theorems 9.5–9.6 hold in uniformly convex metric spaces by routinely modifying the arguments given here. However it does not appear that the same can be said of Theorems 9.7–9.9. In particular, it is not known whether Theorems 9.8 and 9.9 even hold in a uniformly convex Banach space. On the other hand, as we have already noted, Theorem 9.7 holds in any reflexive Banach space but it fails in nonreflexive spaces.*

QUESTION. It remains open whether any of the preceding results extend to spaces of non-positive curvature. There appear to be serious obstacles to carrying out such extensions.

We close this section with a theorem which invokes the Leray–Schauder boundary condition. For this we will use the following continuation principle due to A. Granas.

THEOREM 9.11 ([**89**]). *Let U be a domain (i.e., connected open set) in a complete metric space X, let $f, g : \overline{U} \to X$ be two contraction mappings, and suppose there exists $h : \overline{U} \times [0,1] \to X$ such that*

- (a) $h\left(\cdot, 1\right) = f$, $h\left(\cdot, 0\right) = g$;
- (b) $h\left(x, t\right) \neq x$ *for every* $x \in \partial U$ *and* $t \in [0,1]$;
- (c) *there exists* $k < 1$ *such that* $d\left(h\left(x, t\right), h\left(y, t\right)\right) \leq kd\left(x, y\right)$ *for every* $x, y \in \overline{U}$ *and* $t \in [0,1]$;
- (d) *there exists a constant* $\alpha \geq 0$ *such that for every* $x \in \overline{U}$ *and* $t, s \in [0,1]$,

$$d\left(h\left(x, t\right), h\left(x, s\right)\right) \leq \alpha\left|s - t\right|.$$

Then f has a fixed point if and only if g has a fixed point.

We will also need the following lemma due to Crandall and Pazy.

LEMMA 9.4 ([**56**]). *Let $\{z_n\}$ be a subset of a Hilbert space H and let $\{r_n\}$ be a sequence of positive numbers. Suppose*

$$\langle z_n - z_m, r_n z_n - r_m z_m\rangle \leq 0, \text{ for } m = 1, 2, \cdots.$$

Then if r_n is strictly decreasing $\|z_n\|$ is increasing. If $\|z_n\|$ is bounded, $\lim_{n\to\infty} z_n$ exists.

THEOREM 9.12. *Let U be a bounded connected open set in a complete $CAT(0)$ space X, and suppose $f : \overline{U} \to X$ is nonexpansive. Suppose there exists $p \in U$ such that $x \notin [p, f\left(x\right))$ for all $x \in \partial U$. Then f has a fixed point in $\overline{U}$.*

When X is a Hilbert space, Theorem 9.12 holds under the even weaker assumption that f is a lipschitzian pseudocontractive mapping. This has been known for some time (see [187]). Our proof is patterned after Precup's Hilbert space proof [173] for nonexpansive mappings. We observe here that the CAT(0) inequality is sufficient.

PROOF OF THEOREM 9.12 [120]. Let $t \in (0,1)$ and for $u \in U$ let $f_t(u)$ be the point of the segment $[p, f(u)]$ with distance $td(p, f(u))$ from p. Let $x, y \in U$ and consider the comparison triangle $\bar{\Delta} = \Delta(\bar{p}, \bar{x}, \bar{y})$ of $\Delta(p, x, y)$ in $\mathbb{R}^2$. If $\bar{f}_t(x)$ and $\bar{f}_t(y)$ denote the respective comparison points of $f_t(x)$ and $f_t(y)$ in $\bar{\Delta}$, then by the CAT(0) inequality,

$$d(f_t(x), f_t(y)) \leq \left\| \bar{f}_t(x) - \bar{f}_t(y) \right\| = t \left\| \bar{x} - \bar{y} \right\| = td(x, y).$$

Therefore f_t is a contraction mapping of $U \to X$. Moreover, if $B(p; r) \subset U$, then $f_t : U \to B(p; r)$ for t sufficiently small. Thus f_t has a fixed point for t sufficiently small. Now let $\lambda \in (0,1)$. We apply Theorem 9.11 to show that f_λ has a fixed point. Define the homotopy $h : \bar{U} \times [0,1] \to X$ by setting $h(x, t) = f_{\lambda t}(x)$. Then $h(\cdot, 1) = f_\lambda$ and $h(\cdot, 0)$ is a constant map. If $h(x, t) = x$ for some $x \in \partial U$ and $t \in [0, 1]$, then $f_{\lambda t}(x) = x$ and $x \in [p, f(x)]$. Since this is not possible, condition (b) of Theorem 9.11 holds. Condition (c) holds upon taking k to be λ. Finally,

$$d(h(x, t), h(x, s)) \leq |s - t| d(p, f(x)),$$

for all $t, s \in [0, 1]$, and since U is bounded condition (d) holds. Therefore, by Theorem 9.11, f_λ has a unique fixed point, and it follows that f_t has a unique fixed point x_t for each $t \in (0, 1)$.

Now denote by x_n, $n \in \mathbb{N}$, the point x_t for $t = 1 - 1/n$. For $m, n \in \mathbb{N}$, $m, n > 1$, consider the comparison triangle $\bar{\Delta} = \Delta(0, \bar{f}(x_m), \bar{f}(x_n))$ of $\Delta(p, f(x_m), f(x_n))$ in $\mathbb{R}^2$, and let $\bar{x}_m, \bar{x}_n$ denote the respective comparison points of x_m, x_n. Then, using the fact that f is nonexpansive in conjunction with the CAT(0) inequality,

$$\left\| \bar{f}(x_m) - \bar{f}(x_n) \right\| = d(f(x_n), f(x_m)) \leq d(x_n, x_m) \leq \left\| \bar{x}_n - \bar{x}_m \right\|.$$

Consequently, if $r_m = (m-1)^{-1}$ and $r_n = (n-1)^{-1}$,

$$\langle r_n \bar{x}_n - r_m \bar{x}_m, \bar{x}_n - \bar{x}_m \rangle$$
$$= \langle \bar{f}(x_n) - \bar{f}(x_m), \bar{x}_n - \bar{x}_m \rangle - \left\| \bar{x}_n - \bar{x}_m \right\|^2 \leq 0.$$

Since $\{r_n\}$ is strictly decreasing, $\{\bar{x}_n\}$ converges by Lemma 9.4. Since $d(x_n, x_m) \leq d(\bar{x}_n, \bar{x}_m)$, $\{x_n\}$ converges as well, necessarily to a fixed point of f. $\square$

REMARK 9.4. *It is noteworthy that in the preceding result the domain U is not assumed to be convex nor is there any compactness assumption. See [11] for an interesting extension of Theorem 9.12 to continuous mappings with compact range.*

9.4. A Concept of "Weak" Convergence

In 1976 T.C. Lim [141] introduced a concept of convergence in a general metric space setting which he called strong Δ-convergence. We show here that CAT(0) spaces provide a natural framework for Lim's concept, and that in such a setting Δ-convergence shares many properties of the usual notion of weak convergence in Banach spaces. As a consequence Kirk and Panyanak were able to show in [124] that many Banach space concepts and results which involve weak convergence can be extended to a CAT(0) setting. We discuss those results here. (We should also mention that in [135] T. Kuczumow introduced an identical notion of convergence in Banach spaces, which he calls "almost convergence".)

Throughout, X denotes a complete CAT(0) space. Let $\{x_n\}$ be a bounded sequence in X and for $x \in X$ set

$$r\left(x, \{x_n\}\right) = \limsup_{n \to \infty} d\left(x, x_n\right).$$

The *asymptotic radius* $r\left(\{x_n\}\right)$ of $\{x_n\}$ is given by

$$r\left(\{x_n\}\right) = \inf\left\{r\left(x, \{x_n\}\right) : x \in X\right\},$$

and the *asymptotic center* $A\left(\{x_n\}\right)$ of $\{x_n\}$ is the set

$$A\left(\{x_n\}\right) = \left\{x \in X : r\left(x, \{x_n\}\right) = r\left(\{x_n\}\right)\right\}.$$

It is known [60] that in a CAT(0) space, $A(\{x_n\}))$ consists of exactly one point.

We now turn to the study of Lim's concept in CAT(0) spaces.

DEFINITION 9.5. A sequence $\{x_n\}$ in X is said to Δ-*converge* to $x \in X$ if x is the unique asymptotic center of $\{u_n\}$ for every subsequence $\{u_n\}$ of $\{x_n\}$. In this case we write $\Delta\text{-}\lim_{n \to \infty} x_n = x$ and call x the Δ-limit of $\{x_n\}$.

Next recall that a bounded sequence $\{x_n\}$ in X is said to be *regular* if $r\left(\{x_n\}\right) = r\left(\{u_n\}\right)$ for every subsequence $\{u_n\}$ of $\{x_n\}$. It is known that every bounded sequence in a Banach space has a regular subsequence [87, p. 166]. The proof is metric in nature and carries over to the present setting without change. Since every regular sequence Δ-converges, we see immediately that every bounded sequence in X has a Δ-convergent subsequence.

Notice that given $\{x_n\} \subset X$ such that $\{x_n\}$ Δ-converges to x and given $y \in X$ with $y \neq x$,

$$\limsup_{n \to \infty} d(x_n, x) < \limsup_{n \to \infty} d(x_n, y).$$

Thus X satisfies a condition which is known in Banach space theory as *the Opial property*.

REMARK 9.5. *Every bounded closed convex subset K of X is Δ-closed in the sense that it contains the Δ-limits of all of its Δ-convergent sequences (see Proposition 2.1 in [61]). The following fact is a consequence of this.*

PROPOSITION 9.6. *If a sequence $\{x_n\}$ in X Δ-converges to $x \in X$, then*

$$x \in \bigcap_{k=1}^{\infty} \overline{conv}\{x_k, x_{k+1}, \ldots\},$$

where $\overline{conv}(A) = \bigcap \{B : B \supseteq A \text{ and } B \text{ is closed and convex}\}$.

PROPOSITION 9.7. *Let K be a closed convex subset of X, and let $f : K \to X$ be a nonexpansive mapping. Then the conditions $\{x_n\}$ Δ-converges to x and $d(x_n, f(x_n)) \to 0$, imply $x \in K$ and $f(x) = x$.*

PROOF. Since

$$\limsup_{n \to \infty} d(f(x), x_n) \leq \limsup_{n \to \infty} [d(f(x), f(x_n)) + d(x_n, f(x_n))] = r(x, (x_n)),$$

it must be the case that $f(x) = x$ by uniqueness of asymptotic centers.   □

Notice that Theorem 9.5 is a corollary to the above proposition.

We have seen that CAT(0) spaces satisfy the Opial property. We now show that they also satisfy what is known in Banach space theory as the Kadec–Klee property. For a bounded sequence $\{x_n\}$ in a metric space we denote

$$sep \{x_n\} := \inf \{d(x_n, x_m) : n \neq m\}.$$

THEOREM 9.13. (Kadec–Klee Property) *Let $p \in X$, and let $\varepsilon > 0$. Then there exists $\delta > 0$ such that $d(p, x) \leq 1 - \delta$ for every sequence $\{x_n\} \subset X$ such that $d(p, x_n) \leq 1$, $sep \{x_n\} > \varepsilon$ and $\Delta\text{-}\lim_{n \to \infty} x_n = x$.*

PROOF. For convenience, and without loss of generality, we assume $d(p, x_n) \equiv 1$. By passing to a subsequence if necessary we may suppose $d(x_n, x) \geq \frac{\varepsilon}{2}$ for all n. Let $\triangle(\bar{p}, \bar{x}, \bar{x}_n)$ be a comparison triangle for $\triangle(p, x, x_n)$ in $\mathbb{R}^2$. Then x is the asymptotic center of $\{x_n\}$ relative to the segment $[p, x]$, and $\limsup_n d(x, x_n) = r(\{x_n\})$. For each n, let $\bar{u}_n$ be the point of the segment $[\bar{p}, \bar{x}]$ which is nearest to $\bar{x}_n$, and let u_n be the point of the segment $[p, x]$ for which $d(p, u_n) = d(\bar{p}, \bar{u}_n)$ and $d(u_n, x) = d(\bar{u}_n, \bar{x})$. Let $\theta_n = \measuredangle_{\bar{p}}(\bar{x}, \bar{x}_n)$. By passing to subsequences again we may suppose $\{\bar{u}_n\}$ converges to $\bar{u} \in [\bar{p}, \bar{x}]$, $\{u_n\}$ converges to $u \in [p, x]$, and $\theta_n \to \theta$. Since $d(\bar{x}_n, \bar{x}) = d(x_n, x) \geq \frac{\varepsilon}{2} > 0$ it must be the case that $\theta > 0$. If $d(p, x) = d(\bar{p}, \bar{x}) \leq \cos \theta$, take $\delta = 1 - \cos \theta$. Otherwise $d(p, x) > \cos \theta$ from which $\measuredangle_{\bar{u}_n}(\bar{p}, \bar{x}_n) = \frac{\pi}{2}$ and $d(\bar{p}, \bar{u}_n) = \cos \theta_n$.

This implies $d(\bar{p}, \bar{u}) = \cos \theta$ and $\cos \theta = \lim_{n \to \infty} \cos \theta_n$ can be estimated in terms of ε. In this case, we have (using the CAT(0) inequality),

$$
\begin{aligned}
r(\{x_n\}) &= \limsup_{n \to \infty} d(x, x_n) \\
&= \limsup_{n \to \infty} d(\bar{x}, \bar{x}_n) \\
&\geq \limsup_{n \to \infty} d(\bar{u}_n, \bar{x}_n) \\
&= \limsup_{n \to \infty} d(\bar{u}, \bar{x}_n) \\
&\geq \limsup_{n \to \infty} d(u, x_n).
\end{aligned}
$$

Thus $r(u, \{x_n\}) \leq r(\{x_n\})$. This implies that $u = x$ by uniqueness of the asymptotic center. Hence $\bar{u} = \bar{x}$. But $d(p, u) = d(\bar{p}, \bar{u}) \leq \cos \theta < 1$. We thus conclude that in either case $d(p, u) \leq 1 - \delta$, where δ is positive and depends on ε. $\qquad \square$

9.5. Δ-Convergence of Nets

The notion of Δ-convergence readily extends to nets. We begin by summarizing the results of [125]. A relation $\leq$ is said to be a *partial order* on a set S, and $(S \leq)$ is said to be a *partially ordered set* if for each $a, b, c \in S$

(i) $a \leq a$;
(ii) $a \leq b$ and $b \leq a \Rightarrow a = b$;
(iii) $a \leq b$ and $b \leq c \Rightarrow a \leq c$.

DEFINITION 9.6. A *directed set* is a partially ordered set $(S \leq)$ for which the following condition holds:

(iv) *For each* $a, b \in S$ *there exists* $c \in S$ *such that* $a \leq c$ *and* $b \leq c$.

Recall that a *net* in a set S is a mapping $\phi : I \to S$ where I is a directed set. For $\alpha \in I$ we adopt the notation $\phi = \{x_\alpha\}_{\alpha \in I}$ (and when there is no confusion simply $\{x_\alpha\}$) where it is understood that $\phi(\alpha) = x_\alpha$. If $G \subseteq S$ and $\{x_\alpha\}$ is a net in S, then $\{x_\alpha\}$ is said to be *eventually in* G if there exists $\alpha_0 \in I$ such that $\alpha \geq \alpha_0 \Rightarrow x_\alpha \in G$. If S is a topological space, then the net $\{x_\alpha\}$ is said to converge to $p \in S$ if $\{x_\alpha\}$ is eventually in each neighborhood of p.

DEFINITION 9.7. A net $\{x_\alpha\}$ in a set S is an *ultranet* (*or universal net*) if, given any subset G of S, $\{x_\alpha\}$ is either eventually in G or eventually in $S \backslash G$.

If $\phi : I \to S$ is a net in S and if $\psi : J \to I$, where J is a directed set, then $\phi \circ \psi$ is a *subnet* of ϕ if the following condition holds:

For each $\alpha \in I$ there exists $j_0 \in J$ such that $\psi(j) \geq \alpha$ for all $j \geq j_0$. It is clear from this definition that a subnet of an ultranet is also an ultranet.

The following fact is a remarkable consequence of the Axiom of Choice. See, e.g., [4] for further details.

PROPOSITION 9.8. *Every net in a set S has a subnet which is an ultranet.*

Some other facts are pertinent to our discussion.

PROPOSITION 9.9. *If S is compact in some (Hausdorff) topology, and if $\{x_\alpha\}_{\alpha \in I}$ is an ultranet in S, then $\{x_\alpha\}$ converges to some $p \in S$.*

PROOF. The proof is by contradiction. Suppose not. Then each point $p \in S$ has a neighborhood $U(p)$ such that $\{x_\alpha\}$ is eventually in $S \backslash U(p)$. Since the family $\{U(p)\}_{p \in S}$ is an open cover of the compact set S, there exist $p_1, \cdots, p_n \in S$ such that $S \subseteq \cup_{i=1}^{n} U(p_i)$. For each $i \in \{1, \cdots, n\}$ there exists $\alpha_i \in I$ such that $\alpha \geq \alpha_i \Rightarrow x_\alpha \in S \backslash U(p_i)$. However by (iv) there exists $\alpha \in I$ such that $\alpha \geq \alpha_i$ for $i = 1, \cdots, n$. This implies that x_α does not exist—a contradiction. $\qquad \square$

PROPOSITION 9.10. *Let S_1 and S_2 be sets, and let $\{x_\alpha\}$ be an ultranet in S_1. Then if $f : S_1 \to S_2$ is an arbitrary mapping, $\{f(x_\alpha)\}$ is an ultranet in S_2.*

PROOF. Let $G \subset S_2$ and let

$$f^{-1}(G) = \{x \in S_1 : f(x) \in G\}.$$

Then $\{x_\alpha\}$ is either eventually in $f^{-1}(G)$, in which case $\{f(x_\alpha)\}$ is eventually in G, or $\{x_\alpha\}$ is eventually in $S_1 \backslash f^{-1}(G)$, in which case $\{f(x_\alpha)\}$ is eventually in $S_2 \backslash G$. $\qquad \square$

PROPOSITION 9.11. *Let X be a metric space and let $\{x_\alpha\}$ be a bounded ultranet in X. Then for each $p \in X$, $\{d(x_\alpha, p)\}$ converges.*

PROOF. Define $f : X \to \mathbb{R}$ by setting $f(x) = d(x, p)$. By Proposition 9.10, $\{d(x_\alpha, p)\}$ is an ultranet in a bounded closed subset of $\mathbb{R}$. By Proposition 9.9 $\{d(x_\alpha, p)\}$ converges. $\qquad \square$

We define the notions of asymptotic radius and asymptotic center for net analogous to the way they are defined for sequences. Specifically: Let (X, d) be a metric space and let K be a subset of X. Let I be a directed set, and let $\{x_\alpha\}_{\alpha \in I}$ be a bounded net in X. For $y \in X$, set

$$
\begin{aligned}
r_y(\{x_\alpha\}) &= \lim_\alpha \{\sup \{d(y, x_\beta) : \beta \geq \alpha\}\}; \\
r_K(\{x_\alpha\}) &= \inf \{r_y\{x_\alpha\} : y \in K\}; \\
A_K(\{x_\alpha\}) &= \{x \in K : r_x(\{x_\alpha\}) = r_K(\{x_\alpha\})\}.
\end{aligned}
$$

The number $r_K(\{x_\alpha\})$ is called the *asymptotic radius* of $\{x_\alpha\}$ relative to K and the (possibly empty) set is called the *asymptotic center* of $\{x_\alpha\}$ in K.

A net $\{x_\alpha\}$ is said to be *regular* if each of its subnets has the same asymptotic radius, and $\{x_\alpha\}$ is said to be *uniform* if each of its subnets has the same asymptotic center.

PROPOSITION 9.12 ([**125**]). *Let* $\{x_\alpha\}$ *be a bounded ultranet in a metric space* X. *Then* $\{x_\alpha\}$ *is uniform.*

PROOF. For each $y \in X$ $\{d(y, x_\alpha)\}$ is a bounded ultranet in $\mathbb{R}$. Therefore $\lim_\alpha d(y, x_\alpha) := \varphi(y)$ exists. If $\{x_\beta\}$ is a subnet of $\{x_\alpha\}$, then $\{d(y, x_\beta)\}$ is an ultranet subnet of $\{d(y, x_\alpha)\}$; hence $\lim_\beta d(y, x_\beta) = \varphi(y)$. It follows that every subnet of $\{x_\alpha\}$ has the same asymptotic radius; hence the same asymptotic center. □

Now let K be a closed convex subset of a complete CAT(0) space (X, d). Let $\{x_\alpha\}_{\alpha \in I}$ be a bounded net in K with asymptotic radius r. Then $A_K(\{x_\alpha\})$ consists of exactly one point. To see this, let $\varepsilon > 0$. By assumption there exists $x \in K$ and $\alpha_0 \in I$ such that $d(x, x_\alpha) \leq r + \varepsilon$ for $\alpha \geq \alpha_0$. Therefore

$$C_\varepsilon = \bigcup_{\alpha \in I} \left(\bigcap_{\beta \geq \alpha} B(x_\beta; r + \varepsilon) \right) \neq \emptyset.$$

Since C_ε, being the union of an ascending chain of convex sets, is convex, the closure $\overline{C}_\varepsilon$ of C_ε is also convex. It follows that

$$C := \bigcap_{\varepsilon > 0} \overline{C}_\varepsilon \neq \emptyset.$$

The fact that this intersection consists of a single point follows from the CN inequality (9.1). Specifically, if $u, v \in C$ and $u \neq v$ and if m is the midpoint of the segment $[u, v]$, then by (9.1)

$$d(m, x_\alpha)^2 \leq \frac{d(u, x_\alpha)^2 + d(v, x_\alpha)^2}{2} - \frac{1}{4} d(u, v)^2$$

This implies $r_m(\{x_\alpha\})^2 < r_K(\{x_\alpha\})^2$—a contradiction.

DEFINITION 9.8. Let (X, d) be a complete CAT(0) space. A bounded net $\{x_\alpha\}$ in X is said to Δ-*converge* to $z \in X$ if z is the unique asymptotic center of every subnet of $\{x_\alpha\}$.

Now let $\{x_\alpha\}$ be a bounded net in a complete CAT(0) space. Then $\{x_\alpha\}$ has a subnet which is an ultranet. Since every ultranet is uniform, it Δ-converges to some $z \in X$. Thus we have the following:

PROPOSITION 9.13. *Every bounded net in a complete CAT(0) space has a Δ-convergent subnet.*

The preceding fact can be reformulated as follows (cf., Theorem 3 of [141]).

PROPOSITION 9.14. *Every bounded closed convex set in a complete CAT(0) is Δ-compact.*

9.6. A Four Point Condition

In this section (X, d) always denotes a complete CAT(0) space, and we assume that X satisfies the following seemingly mild geometric condition.

(Q_4) For points $x, y, p, q \in X$,

$$\left. \begin{array}{l} d(x, p) < d(x, q) \\ d(y, p) < d(y, q) \end{array} \right\} \Rightarrow d(m, p) \leq d(m, q)$$

for any point m on the segment $[x, y]$.

This condition was introduced in [124], and it is easy to see that it holds in many CAT(0) spaces, including Hilbert spaces and $\mathbb{R}$-trees. This condition has been studied more deeply by Espínola and Fernández-León in [74]. They show in particular that any CAT(0) space of constant curvature satisfies (Q_4), but any CAT(0) gluing of two such spaces of different constant curvature fails the (Q_4) condition.

As we observed above (Proposition 9.6), if a sequence $\{x_n\}$ in X Δ-converges to $x \in X$, then

$$x \in \bigcap_{k=1}^{\infty} \overline{conv}\{x_k, x_{k+1}, \ldots\},$$

and, as is the case for weak convergence in a Banach space, it is natural to ask when

$$\{x\} = \bigcap_{k=1}^{\infty} \overline{conv}\{x_k, x_{k+1}, \ldots\}.$$

A positive answer is given by Ahmadi Kakavandi in [1], where it is shown that the following strengthening of condition (Q_4) is sufficient.

$\overline{(Q_4)}$ For points $x, y, p, q \in X$,

$$\left. \begin{array}{l} d(x, p) \leq d(x, q) \\ d(y, p) \leq d(y, q) \end{array} \right\} \Rightarrow d(m, p) \leq d(m, q)$$

for any point m on the segment $[x, y]$.

We now discuss another ultrapower technique. (See Chap. 7 for a previous discussion.) Assume that K is a bounded closed convex subset of a complete CAT(0) space X. Let $\mathcal{U}$ be a nontrivial ultrafilter on the natural numbers $\mathbb{N}$. Fix $p \in X$, and let $\tilde{X}_{\mathcal{U}}$ denote the metric space ultrapower of X

over $\mathcal{U}$ relative to p. In this case the elements of $\tilde{X}_{\mathcal{U}}$ consist of equivalence classes $\tilde{x} := [(x_i)]_{i \in \mathbb{N}}$ for which

$$\lim_{\mathcal{U}} d(x_i, p) < \infty,$$

with $(u_i) \in [(x_i)]$ if and only if $\lim_{\mathcal{U}} d(x_i, u_i) = 0$. It is known that $\tilde{X}_{\mathcal{U}}$ is also a CAT(0) space [**36**, p. 187]. We use $\dot{x}$ to denote the class $[(x_i)]$ with $x_i \equiv x$, and $\dot{X}$ denotes the canonical isometric embedding of X in $\tilde{X}_{\mathcal{U}}$.

The following ultrapower characterization of Δ-convergence can be found in [**61**].

PROPOSITION 9.15. *A regular sequence* $(x_n) \subset X$ Δ-*converges to* $x \in X$ *if and only if for any nontrivial ultrafilter* $\mathcal{U}$ *over* $\mathbb{N}$, $\dot{x}$ *is the unique point of* $\dot{X}$ *which is nearest to* $\tilde{x} := [(x_n)]$ *in the ultrapower* $\tilde{X}_{\mathcal{U}}$.

PROOF. ($\Rightarrow$) Suppose x is the asymptotic center of $\{x_n\}$, and suppose $d_{\mathcal{U}}(\dot{y}, \tilde{x}) \leq d_{\mathcal{U}}(\dot{x}, \tilde{x})$ for some $y \in X$. Choose a subsequence $\{u_n\}$ of $\{x_n\}$ such that

$$\lim_{n \to \infty} d(y, u_n) = \liminf_{n \to \infty} d(y, x_n).$$

Using the fact that $\{x_n\}$ is regular we have

$$\begin{aligned}
\lim_{n \to \infty} d(y, u_n) &\leq \lim_{\mathcal{U}} d(y, x_n) \\
&= d_{\mathcal{U}}(\dot{y}, \tilde{x}) \\
&\leq d_{\mathcal{U}}(\dot{x}, \tilde{x}) \\
&\leq \limsup_{n \to \infty} d(x, x_n) \\
&= r(\{x_n\}) \\
&= \limsup_{n \to \infty} d(x, u_n).
\end{aligned}$$

Thus $\lim_{n \to \infty} d(y, u_n) \leq \limsup_{n \to \infty} d(x, u_n)$, and $y = x$ by uniqueness of the asymptotic center.

($\Leftarrow$) Suppose $\dot{x}$ is the unique point of $\dot{X}$ which is nearest to $\tilde{x} := [(x_n)]$, and suppose y is the asymptotic center of $\{x_n\}$. Then by the implication ($\Rightarrow$) $\dot{y}$ is the unique point of $\dot{X}$ which is nearest to $\tilde{x}$, whence $\dot{x} = \dot{y}$; thus $x = y$. $\square$

PROPOSITION 9.16. *Suppose* X *satisfies* (Q_4), *and suppose* $\{x_n\}$ *and* $\{y_n\}$ *both* Δ-*converge to* $p \in X$. *Suppose* $m_n \in [x_n, y_n]$ *satisfies* $d(x_n, m_n) = \lambda d(x_n, y_n)$ *for fixed* $\lambda \in (0, 1)$. *Then* $\{m_n\}$ *also* Δ-*converges to* p.

PROOF. We pass to the ultrapower $\tilde{X}_{\mathcal{U}}$ of Proposition 9.15. Thus $\dot{p}$ is the unique point of $\dot{X}$ which is nearest to both $\tilde{x}$ and $\tilde{y}$. Then some subsequence of $\{m_n\}$, which we again denote by $\{m_n\}$, Δ-converges to q, and $\dot{q}$ is the unique point of $\dot{X}$ which is nearest to $\tilde{m}$. We pass to corresponding subsequences of $\{x_n\}$ and $\{y_n\}$ and relabel as at the outset. Assume $\dot{q} \neq \dot{p}$. Then $d_{\mathcal{U}}(\tilde{x}, \dot{p}) < d_{\mathcal{U}}(\tilde{x}, \dot{q})$ and $d_{\mathcal{U}}(\tilde{y}, \dot{p}) < d_{\mathcal{U}}(\tilde{y}, \dot{q})$, while $d_{\mathcal{U}}(\tilde{m}, \dot{q}) < d_{\mathcal{U}}(\tilde{m}, \dot{p})$.

It follows that one can choose n so that $d(x_n, p) < d(x_n, q)$, $d(y_n, p) < d(y_n, q)$, and $d(m_n, q) < d(m_n, p)$. This contradicts condition (Q_4). Thus every subsequence of the original sequence $\{m_n\}$ Δ-converges to p, and so $\{m_n\}$ itself Δ-converges to p. $\qquad\square$

It is also shown in [124] that the approach of Kirk–Sims in [131] carries over to CAT(0) spaces which satisfy (Q_4), provided one makes certain minor adjustments. If K is a closed convex subset of a Banach space X, a continuous mapping $f : K \to X$ is said to be *locally almost nonexpansive* (LANE) if for each $x \in K$ and $\varepsilon > 0$ there exists a weak neighborhood U_x of x such that for $u, v \in U_x$, $\|f(u) - f(v)\| \leq \|u - v\| + \varepsilon$. The concept is due to R. Nussbaum [162]. He proved that if X is uniformly convex and if $f : K \to X$ is a LANE mapping, then $I - f$ is demiclosed on K, in the sense that the conditions $\{x_n\} \subset K$ converges weakly to $x \in X$ and $\lim_{n\to\infty} \|(I - f)(x_n) - y\| = 0 \Rightarrow x \in K$ and $x - f(x) = y$.

It is not even possible to formulate the above result, as stated, in a CAT(0) setting. However it is possible to use the notion of Δ-convergence to formulate a precise analogue.

DEFINITION 9.9. Let K be a closed convex subset of a complete CAT(0) space. A continuous mapping $f : K \to X$ is said to be *locally almost nonexpansive* (*LANE*) if for each $x \in K$ and $\varepsilon > 0$ the following condition holds: If $\{u_n\}, \{v_n\}$ are two sequences in K which Δ-converge to x, then there exists $N \in \mathbb{N}$ such that

$$(9.4) \qquad d(f(u_n), f(v_n)) \leq d(u_n, v_n) + \varepsilon \quad \text{whenever } n \geq N.$$

It is now possible to follow the approach of [131]. Let

$$\tilde{K}_{\mathcal{U}} = \{\tilde{x} = [(x_i)] : x_i \in K \text{ for each } i\}.$$

Assume $f : K \to X$ is a LANE mapping. For $\tilde{x} = [(x_i)] \in \tilde{K}_{\mathcal{U}}$, define $\tilde{f} : \tilde{K}_{\mathcal{U}} \to \tilde{X}_{\mathcal{U}}$ by setting

$$\tilde{f}(\tilde{x}) = [(f(x_i))].$$

For each $x \in K$, let

$$W_x = \left\{ \tilde{x} = [(x_i)] \in \tilde{K}_{\mathcal{U}} : \Delta\text{-}\lim_{\mathcal{U}} x_i = x \right\}.$$

If $\{x_n\}$ Δ-converges to x and $\{y_n\}$ Δ-converges to y, then $\dot{x}$ and $\dot{y}$ are the unique points of $\dot{X}$ which are nearest $\tilde{x}$ and $\tilde{y}$, respectively. Since the nearest point projection from $\tilde{X}_{\mathcal{U}}$ onto $\dot{X}$ is nonexpansive by Proposition 9.5, $d(x, y) = d_{\mathcal{U}}(\dot{x}, \dot{y}) \leq d_{\mathcal{U}}(\tilde{x}, \tilde{y})$. It follows that the sets W_x are closed, and by Proposition 9.16 they are also convex. The remaining details follow as in [131] and lead to the following analog of Proposition 9.7.

THEOREM 9.14 (cf., [203]). *Let K be a closed convex subset of a complete CAT(0) space X, suppose X satisfies (Q_4), and let $f : K \to X$ be a continuous LANE mapping. Then the conditions $\{x_n\}$ Δ-converges to x and*

$$\lim_{n \to \infty} d(x_n, f(x_n)) = 0$$

imply $x \in K$ and $f(x) = x$.

A question posed in [124] is whether every CAT(0) space satisfies (Q_4). This question is answered negatively in [74]. Among other things it is also shown that Proposition 9.16 holds under a somewhat weaker assumption than (Q_4), a fact which answers negatively another question posed in [124].

9.7. Multimaps and Invariant Approximations

Let (X, d) be a metric space, let 2^X be the family of all subsets of X and let $CB(X)$ be the family of nonempty bounded closed subsets of X. For $A \in CB(X)$ and $\varepsilon > 0$, let

$$N_\varepsilon(A) = \{y \in X : dist(y, A) \leq \varepsilon\}.$$

The *Hausdorff–Pompeiu metric* H on $CB(X)$ is defined as follows:

$$H(A, B) = \max \left\{ \sup_{a \in A} dist(a, B), \sup_{b \in B} dist(b, A) \right\}.$$

This can be written in a more geometrical form as follows:

$$H(A, B) = \inf \{\varepsilon > 0 : A \subseteq N_\varepsilon(B) \text{ and } B \subseteq N_\varepsilon(A)\}.$$

A set-valued mapping $T : X \to CB(X)$ is said to be *nonexpansive* if

$$H(T(x), T(y)) \leq d(x, y) \text{ for all } x, y \in X.$$

The following is the fundamental fixed point theorem for set-valued mappings in a CAT(0) space.

THEOREM 9.15. *Suppose (X, d) is a complete CAT(0), let K be a bounded closed convex subset of X, and suppose $T : K \to 2^K$ is a nonexpansive set-valued mapping whose values are nonempty compact subsets of K. Then T has a fixed point.*

PROOF. Since asymptotic centers of bounded sequences are unique in CAT(0) spaces, it is possible to follow the standard proof of the Banach space analog of this theorem in a uniformly convex space (cf., [87, p. 165]). □

As an application of this theorem we obtain the following results of Shahzad and Markin [199]. In this result the mappings $f : X \to X$ and $T : X \to 2^X$ are said to *commute* if for all $x \in X$, $T(x) \neq \emptyset$ and $f(T(x)) \subset T(f(x))$.

THEOREM 9.16. *Let (X,d) be a complete bounded CAT(0) space and let $f : X \to X$ be nonexpansive. Suppose $T : X \to 2^X$ is a nonexpansive mapping whose values are compact and convex. If the mappings f and T commute, then there exists $x^* \in X$ such that $x^* = f(x^*) \in T(x^*)$.*

PROOF. By Theorem 9.4, f has a nonempty fixed point set A in X which is closed and convex. Since f and T commute, $T(x)$ is invariant under f for all $x \in A$. Since $T(x)$ is also closed bounded and convex, it is also a CAT(0) space, so again by Theorem 9.4, f has a fixed point in $T(x)$. Thus $T(x) \cap A \neq \emptyset$ for each $x \in A$. Now consider the mapping $T' : A \to 2^A$ defined by $T'(x) = T(x) \cap A$, $x \in A$. We claim that T' is nonexpansive. Indeed, if $u \in T'(x)$ for some $x \in A$, let v be the unique closest point to u in $T(y)$ for some $y \in A$. Then $d(u,v) = \inf_{w \in T(y)} d(u,w)$. However $d(u,f(v)) = d(f(u),f(v)) \leq d(u,v)$. Since v is the unique closest point to u in $T(y)$ and since $f(v) \in T(y)$ it must be the case that $v = f(v)$. Therefore $v \in T(y) \cap A = T'(y)$. Since the argument is symmetric in x and y, it follows that

$$H(T'(x),T'(y)) \leq H(T(x),T(y)) \leq d(x,y) \text{ for all } x,y \in A.$$

By Theorem 9.15, T' has a fixed point x^*. Thus $x^* \in T'(x^*) = T(x^*) \cap A$, whence $x^* = f(x^*) \in T(x^*)$.   □

THEOREM 9.17. *Suppose K is a closed bounded convex subset of a complete CAT(0) space (X,d). Suppose $f : K \to K$ and $T : K \to 2^X$ are nonexpansive with T taking compact convex values, and suppose $T(x) \cap K \neq \emptyset$ for each $x \in K$. If the mappings f and T commute (i.e., $f(T(x) \cap K) \subset T(f(x)) \cap K$ for all $x \in K$), then there exists $x^* \in X$ such that $x^* = f(x^*) \in T(x^*)$.*

PROOF. As in the previous theorem the mapping f has a nonempty closed convex fixed point set A in K. By the definition of commuting mappings, $f(y) \in T(f(x)) \cap K = T(x) \cap K$ for $y \in T(x) \cap K$ and $x \in A$, and therefore $T(x) \cap K$ is invariant under f for each $x \in A$. It follows that f has a fixed point in $T(x) \cap K$, so $T(x) \cap A \neq \emptyset$ for each $x \in A$. Now define $T'(x) = T(x) \cap A$, $x \in A$ and complete the proof as in Theorem 9.16.   □

The following is one of the main results of Shahzad [198]. Let $\partial_K C$ denote the relative boundary of $C \subset K$ with respect to K.

THEOREM 9.18. *Let K be a closed bounded convex subset of a complete CAT(0) space X and f a nonexpansive self-mapping of K. Then for any closed convex subset C of K such that $f(\partial_K C) \subset C$ we have $P_{fix(f)}(C) \subset C$.*

PROOF. Fix $u \in C$, and define the mapping $f_t : Y \to K$ by taking $f_t(x)$ to be the point of $[u, f(x)]$ at distance $td(u, f(x))$ from u. Then by convexity of the metric

$$d(f_t(x), f_t(y)) \leq td(x,y)$$

for all $x, y \in C$. This shows that $f_t : C \to K$ is a contraction. Let P_C be the proximinal nonexpansive retraction of K into C. Then $P_C f_t$ is a contraction self-mapping of C. By the Banach Contraction Principle, there exists a unique fixed point $y_t \in C$ of $P_C f_t$. Thus

$$d(f_t y_t, y_t) = \inf\{d(f_t y_t, z) : z \in C\}.$$

Since $f(\partial_K C) \subset C$, we have $f_t(\partial_K C) \subset C$ and so we have $f_t(y_t) = y_t \in [u, f(y_t)]$. Note that $A = Fix(f)$ is nonempty closed bounded convex by Theorem 9.4. Now Theorem 9.8 guarantees that $\lim_{t \to 1^-} y_t$ converges to the unique fixed point of f which is nearest u. As a result, $\lim_{t \to 1^-} y_t = P_A(u) \in C$. Since X is a CAT(0) space, P_A is nonexpansive and $P_A(C) \subset C$.

$\square$

REMARK 9.6. *Let K be a closed bounded convex subset of a complete CAT(0) space X and $f : K \to X$ a nonexpansive mapping. Then there exists an element $x^* \in K$ such that*

$$d(f(x^*), x^*) = d(f(x^*), K).$$

To see this, let P_K be the proximinal nonexpansive retraction of X into K. Then $P_K \circ f$ is a nonexpansive self-mapping of K and so has a fixed point x^*. Hence

$$d(f(x^*), x^*) = d(f(x^*), K).$$

The following is one of the invariant approximation results of Shahzad and Markin [199].

THEOREM 9.19. *Suppose K is a closed convex subset of a complete CAT(0) space (X, d) with $int(K)$ nonempty. Suppose $f : X \to X$ and $T : X \to 2^X$ are nonexpansive mappings with T taking compact convex values, and suppose $f(\partial K) \subset K$ and $T(\partial K) \subset K$. If the mappings f and T commute and $y \in fix(f) \cap fix(T)$, then there exists $x^* \in X$ such that $d(y, x^*) = d(y, K)$ and $x^* = f(x^*) \in T(x^*)$.*

PROOF. Let $P_K(y) = B(y; d(y, K)) \cap K = x^*$ be the unique closest point to $y \in K$. Then the point x^* must lie in $\partial(K)$. Otherwise, if $x^* \in int(K)$, for $\varepsilon > 0$ sufficiently small there is a ball $B(x^*; \varepsilon)$ that is contained in $int(K)$. Since $B(x^*; \varepsilon)$ is a closed and convex set, let w denote the unique closest point in $B(x^*; \varepsilon)$ to y. Thus $d(y, w) < d(y, x^*)$, which is a contradiction. Since $d(y, f(x^*)) \leq d(y, x^*)$ and $f(x^*) \in K$, the uniqueness of x^* implies $x^* = f(x^*)$. Again, since $y \in T(y)$, we have $B(y; d(y, K)) \cap T(x^*) \neq \emptyset$. Since $T(x^*) \subset K$, this implies $x^* \in T(x^*)$. Thus x^* is the required point. $\square$

We now turn to an extension of Theorem 9.15. For convenience and brevity we again work in an ultrapower setting and follow the approach of [61]. Assume that K is a bounded closed convex subset of a complete CAT(0) space X. Let $\mathcal{U}$ be a nontrivial ultrafilter on the natural numbers $\mathbb{N}$. Fix $p \in X$, and let $\tilde{X}_\mathcal{U}$ denote the metric space ultrapower of X over $\mathcal{U}$ relative to p. A nonexpansive set-valued mapping $T : K \to \mathcal{CB}(X)$ induces a nonexpansive set-valued mapping $\tilde{T}$ defined on $\tilde{K}$ as follows:

$$\tilde{T}(\tilde{x}) = \left\{ \tilde{u} \in \tilde{X}_\mathcal{U} : \exists \text{ a representative } (u_n) \text{ of } \tilde{u} \text{ with } u_n \in T(x_n) \text{ for each } n \right\}.$$

To see that $\tilde{T}$ is nonexpansive (and hence well-defined), let $\tilde{x}, \tilde{y} \in \tilde{K}_\mathcal{U}$, with $\tilde{x} = [(x_n)]$ and $\tilde{y} = [(y_n)]$. Then

$$H\left(\tilde{T}(\tilde{x}), \tilde{T}(\tilde{y})\right) \leq \lim_\mathcal{U} H\left(T(x_n), T(y_n)\right)$$
$$\leq \lim_\mathcal{U} d(x_n, y_n)$$
$$= d_\mathcal{U}(\tilde{x}, \tilde{y}).$$

The following fact will be needed (see, e.g., [**105**]).

(9.5)   If $S \subseteq K$ is compact, then $\dot{S} = \tilde{S}$.

THEOREM 9.20. *Let K be a closed convex subset of a complete $CAT(0)$ space X, and let $T : K \to 2^X$ be a nonexpansive mapping whose values are nonempty compact subsets of X. Suppose $dist(x_n, T(x_n)) \to 0$ for some bounded sequence $\{x_n\} \subset K$. Then T has a fixed point.*

PROOF. By passing to a subsequence we may suppose $\{x_n\}$ is regular and hence Δ-converges to some point $x \in X$. By Proposition 9.15 $\dot{x}$ is the unique point of X which is nearest to $\tilde{x} := [(x_n)]$. Since $x \in K$, $\dot{x} \in \dot{K}$. Also, $\tilde{x}$ must lie in an r-neighborhood of $\tilde{T}(\dot{x})$ for $r = H\left(\tilde{T}(\tilde{x}), \tilde{T}(\dot{x})\right)$. Since $\tilde{T}(\dot{x})$ is compact, $dist\left(\tilde{x}, \tilde{T}(\dot{x})\right) = d_\mathcal{U}(\tilde{x}, \dot{u})$ for some $\dot{u} \in \tilde{T}(\dot{x})$. But since $\tilde{T}(\dot{x}) \subset \dot{X}$, if $\dot{u} \neq \dot{x}$ we have the contradiction

$$d_\mathcal{U}(\tilde{x}, \dot{u}) > d_\mathcal{U}(\tilde{x}, \dot{x}) \geq H\left(\tilde{T}(\tilde{x}), \tilde{T}(\dot{x})\right) = r.$$

Therefore $\dot{x} = \dot{u} \in \tilde{T}(\dot{x})$. However $\tilde{T}(\dot{x}) = \widetilde{T(x)}$, so by (9.5) this in turn implies $x \in T(x)$. $\square$

REMARK 9.7. *Convexity of K is needed in the preceding argument only to assure that the asymptotic center of $\{x_n\}$ lies in K. The theorem actually holds under the weaker assumption that K is closed and contains the asymptotic centers of all of its regular sequences.*

9.8. Quasilinearization

Let (X, d) be a metric space. Berg and Nikolaev in [26] introduced the concept of *quasilinearization* in metric spaces. Formally denote a pair (a, b) in $X \times X$ by $\overrightarrow{ab}$ and call it a vector. Quasilinearization is defined as a map $\langle \cdot, \cdot \rangle : X \times X \to X \times X \to \mathbb{R}$ defined by

$$(9.6) \qquad \langle \overrightarrow{ab}, \overrightarrow{cd} \rangle = \frac{1}{2} \left(d^2(a, d) + d^2(b, c) - d^2(a, c) - d^2(b, d) \right).$$

It is easily seen that $\langle \overrightarrow{ab}, \overrightarrow{cd} \rangle = \langle \overrightarrow{cd}, \overrightarrow{ab} \rangle$, $\langle \overrightarrow{ab}, \overrightarrow{cd} \rangle = -\langle \overrightarrow{ba}, \overrightarrow{cd} \rangle$ and

$$\langle \overrightarrow{ax}, \overrightarrow{cd} \rangle + \langle \overrightarrow{xb}, \overrightarrow{cd} \rangle = \langle \overrightarrow{ab}, \overrightarrow{cd} \rangle$$

for all $a, b, c, d, x \in X$, because

$$\langle \overrightarrow{ax}, \overrightarrow{cd} \rangle = \frac{1}{2} \left(d^2(a, d) + d^2(x, c) - d^2(a, c) - d^2(x, d) \right)$$

and

$$\langle \overrightarrow{xb}, \overrightarrow{cd} \rangle = \frac{1}{2} \left(d^2(x, d) + d^2(b, c) - d^2(x, c) - d^2(b, d) \right).$$

It is known [26] that a geodesically connected metric space is a CAT(0) space if and only if it satisfies the Cauchy–Schwarz inequality:

$$\langle \overrightarrow{ab}, \overrightarrow{cd} \rangle \leq d(a, b) \, d(c, d).$$

Using the concept of quasilinearization, the authors in [2] introduce another notion of "weak" convergence in complete CAT(0) spaces.

DEFINITION 9.10. Let (X, d) be a complete CAT(0) space. A sequence $\{x_n\}$ in X is said to *w-converge to* $x \in X$ if for each $y \in X$, $\lim_{n \to \infty} \langle \overrightarrow{xx_n}, \overrightarrow{xy} \rangle = 0$.

It is obvious that convergence in the metric implies w-convergence, and it is easy to check that w-convergence implies Δ-convergence (see Proposition 2.5 in [2]). It is shown in [1] that the converse is not true. However the following result provides an explicit relationship between w-convergence and Δ-convergence.

THEOREM 9.21 ([1]). *Let (X, d) be a complete CAT(0) space. Then a sequence $\{x_n\}$ in X Δ-converges to $x \in X$ if and only if for each $y \in X$,*

$$\limsup_{n \to \infty} \langle \overrightarrow{xx_n}, \overrightarrow{xy} \rangle \leq 0.$$

REMARK 9.8. Let (X, d) be a metric space. One can define a weaker topology $\mathfrak{T}_w$ on X by taking $\mathfrak{T}_w$ to be the weakest topology on X for which the function $z \mapsto d(x, z) - d(y, z)$ is continuous, and $\mathfrak{T}_c$ to be the weakest topology on X for which metrically closed sets are $\mathfrak{T}_c$-closed. In view of the following result, it appears that the latter definition has greater relevance for CAT(0) spaces. N. Monod has made the following observation in [157].

THEOREM 9.22. *Let (X, d) be a complete $CAT(0)$ space, and $K \subseteq X$ a bounded closed convex set. Then K is compact in the topology $\mathfrak{T}_c$.*

Finally we call attention to the following result of Dehghan and Rooin [59]. In this theorem P_K denotes the nearest point projection of X onto K.

THEOREM 9.23. *Let K be a convex subset of a $CAT(0)$ space X. Let $x \in X$ and $y \in K$. Then $y = P_K(x)$ if and only if*

$$\langle \overrightarrow{xy}, \overrightarrow{yu} \rangle \geq 0 \ \text{for all } u \in K.$$

CHAPTER 10

Ptolemaic Spaces

A metric space (X, d) is said to be *ptolemaic* if it satisfies the Ptolemy inequality:

$$d(x, y)\, d(z, p) \quad \leq \quad d(x, z)\, d(y, p) + d(x, p)\, d(y, z)$$
$$\text{for all } x, y, z, p \quad \in \quad X.$$

It is known [196] that a normed space is an inner product space if and only if it is ptolemaic. Also, for each normed space $(X, \|\cdot\|)$ there is a constant $C \in [1, 2]$ such that

$$\|x - y\|\, \|z - p\| \leq C\, (\|x - z\|\, \|y - p\| + \|x - p\|\, \|y - z\|)$$

for all $x, y, z, p \in X$.

The smallest constant $C_p(X)$ for which the above inequality holds is called the *Ptolemy constant* of the space X. Among other things, it is known that if X is a Banach space for which $C_p(X) < (1 + \sqrt{3})/2$, then X has uniform normal structure.

In is also known [44] that CAT(0) spaces are ptolemaic. However a geodesic ptolemaic space is not necessarily a CAT(0) space. In fact such spaces need not even be uniquely geodesic, hence not necessarily Busemann spaces. On the other hand, a metric space is CAT(0) if and only if it is ptolemaic and Busemann convex (see Foertsch et al. [82]).

The metric of a ptolemaic geodesic space is always convex. To see this, let $u = (1 - t)\, y + tz$. Then $d(u, y) = td(y, z)$ and $d(u, z) = (1 - t)\, d(y, z)$. Now apply the ptolemaic inequality as follows:

$$d(x, (1 - t)\, y + tz)\, d(y, z) \leq d(x, y)\, d((1 - t)\, y + tz, z)$$
$$+ d(x, z)\, d((1 - t)\, y + tz, y) = (1 - t)\, d(x, y)\, d(y, z) + td(x, z)\, d(y, z).$$

We say that X admits a *continuous midpoint map* if there exists an $m : X \times X \to X$ such that

$$d(x, m(x, y)) = d(y, m(x, y)) = \frac{d(x, y)}{2}$$

and for $x, y \in X$, the conditions $x_n \to x$ and $y_n \to y$ imply $m(x_n, y_n) \to m(x, y)$. It is also shown in [82] that a ptolemaic geodesic space with a continuous midpoint map is uniquely geodesic.

© Springer International Publishing Switzerland 2014   95
W. Kirk, N. Shahzad, *Fixed Point Theory in Distance Spaces*,
DOI 10.1007/978-3-319-10927-5_10

10.1. Some Properties of Ptolemaic Geodesic Spaces

THEOREM 10.1. ([83]) *Let X be a ptolemaic geodesic space with a continuous midpoint map. Then X is strictly convex.*

DEFINITION 10.1. Let X be a geodesic space. We say that X *admits a uniformly continuous midpoint map* if there exists a map $m : X \times X \to X$ such that

$$d\left(x, m\left(x, y\right)\right) = d\left(y, m\left(x, y\right)\right) = \frac{d\left(x, y\right)}{2} \text{ for all } x, y \in X,$$

and for $n \in \mathbb{N}$ and the conditions $x_n, x'_n, y_n, y'_n \in X$ with $\lim_{n \to \infty} d\left(x_n, x'_n\right) = 0$ and $\lim_{n \to \infty} d\left(y_n, y'_n\right) = 0$ imply $\lim_{n \to \infty} d\left(m\left(x_n, y_n\right), m\left(x'_n, y'_n\right)\right) = 0$.

Every Busemann space admits a uniformly continuous midpoint map, but the converse is not true (see [75]).

A geodesic space (X, d) is said to be *reflexive* if every descending sequence of nonempty bounded closed convex subsets of X has nonempty intersection. The following is the main result of [75].

THEOREM 10.2. *A complete geodesic ptolemaic space (X, d) with a uniformly continuous midpoint map is reflexive.*

As a consequence of the above, if K is a nonempty, bounded, closed, and convex subset of X, then every nonexpansive mapping $T : K \to K$ has a nonempty closed and convex fixed point set.

DEFINITION 10.2. A geodesic metric space (X, d) is said to be *uniformly convex* if for any $r > 0$ and any $\varepsilon \in (0, 2]$, there exists $\delta \in (0, 1]$ such that for all $a, x, y \in X$ with $d\left(x, a\right) \leq r$, $d\left(y, a\right) \leq r$ and $d\left(x, y\right) \geq \varepsilon r$ it is the case that

$$d\left(m, a\right) \leq \left(1 - \delta\right) r,$$

where m denotes the midpoint of any geodesic segment $[x, y]$.

QUESTION. *A natural question to raise at this point is whether a complete geodesic ptolemaic space with a uniformly continuous midpoint map is a uniformly convex metric space.*

The following is Theorem 1.1 of [82].

THEOREM 10.3. *Let X be an arbitrary Ptolemy space. Then X can be isometrically embedded into a complete geodesic Ptolemy space $\hat{X}$.*

PROOF. ([82]) Explicitly construct the complete geodesic Ptolemy metric space $\hat{X}$ as follows. First, add midpoints to X in order to obtain a Ptolemy metric space $\mathfrak{U}(X)$ which has the midpoint property. Then pass to an ultraproduct of $\mathfrak{U}(X)$.

Let Σ denote the set of unordered tuples in X. Formally,

$$\Sigma = \{\{x_1, x_2\} \subset X : x_1, x_2 \in X\},$$

that is Σ consists of all subsets of X with one or two elements.

On Σ define a metric via

$$d(\{x_1, x_2\}, \{y_1, y_2\})$$
$$= \begin{cases} \frac{1}{4}[d(x_1, y_1) + d(x_1, y_2) + d(x_2, y_1) + d(x_2, y_2)] & \text{if } \{x_1, x_2\} \neq \{y_1, y_2\} \\ 0 & \text{otherwise} \end{cases}$$

for all $\{x_1, x_2\}, \{y_1, y_2\} \in \Sigma$. This indeed defines a metric on Σ. In order to verify this, one has to prove the triangle inequality

$$d(\{x_1, x_2\}, \{y_1, y_2\}) \leq d(\{x_1, x_2\}, \{z_1, z_2\}) + d(\{z_1, z_2\}, \{y_1, y_2\})$$

for all $\{x_1, x_2\}, \{y_1, y_2\}, \{z_1, z_2\} \in \Sigma$. If two of the triples coincide, the validity of the inequality is evident, and otherwise it just follows by repeated application of the triangle inequality:

$$\frac{1}{4}[d(x_1, y_1) + d(x_1, y_2) + d(x_2, y_1) + d(x_2, y_2)]$$
$$\leq \frac{1}{4}[d(x_1, z_1) + d(z_1, y_1) + d(x_1, z_2) + d(z_2, y_2)$$
$$+ d(x_2, z_1) + d(z_1, y_1) + d(x_2, z_2) + d(z_2, y_2)]$$

At this point notice that if $x, y \in X$, $x \neq y$, then $\{x, y\}$ is a midpoint of x and y in $M(X)$ because

$$d(\{x, x\}, \{x, y\}) = \frac{1}{4}[d(x, x) + d(x, y) + d(x, x) + d(x, y)]$$
$$= \frac{1}{2}d(x, y)$$
$$= d(\{y, y\}, \{x, y\}).$$

Moreover it is asserted in [82] that the space $M(X) := (\Sigma, d)$ is Ptolemy, that is, it satisfies

$$d(\{x_1, x_2\}, \{y_1, y_2\})\, d(\{z_1, z_2\}, \{u_1, u_2\})$$
$$\leq d(\{x_1, x_2\}, \{z_1, z_2\})\, d(\{y_1, y_2\}, \{u_1, u_2\})$$
$$+ d(\{x_1, x_2\}, \{u_1, u_2\})\, d(\{y_1, y_2\}, \{z_1, z_2\}).$$

Once again, the validity of this inequality is evident if two of the tuples coincide, and otherwise it follows by applying the Ptolemy inequality in X 16 times. [We take the authors' word for this.]

Note further that X isometrically embeds into $M(X)$ via $x \mapsto \{x, x\}$. Thus it is possible to identify X with a subset of $M(X)$.

Now define $M^0(X) := X$, and $M^{n+1}(X) := M(M^n(X))$, and set $\mathfrak{U}(X) = \bigcup_{n=0}^{\infty} M^n(X)$. From the above, this space is a Ptolemy metric space. Moreover, it has the midpoint property. Namely, each pair $x, y \in \mathfrak{U}(X)$ is in some $M^n(X)$ and $\{x, y\} \in M^{n+1}(X)$ is a midpoint of x and y. By passing to an ultraproduct $\hat{X}$ of $\mathfrak{U}(X)$ over some nontrivial ultrafilter $\mathcal{U}$, one obtains a complete Ptolemy metric space which has the midpoint property. By Menger's Theorem, $\hat{X}$ is a geodesic space. $\square$

10.2. Another Four Point Condition

Here is a quote from Foertsch et al. [82].

> Finally, we want to draw the reader's attention to a recent joint work of Berg and Nikolaev. In [26] the authors consider another four point condition, which one derives from the Ptolemy inequality by replacing the products of distances through the sums of their squares. Especially in light of our Theorem 1.1, it seems remarkable to us, that such a variant of the Ptolemy inequality indeed forces a geodesic space to be CAT(0).

The four point condition referred above is the following *quadrilateral inequality condition.*

$$d^2(x,y) + d^2(z,p)$$
$$(10.1) \qquad \leq \quad d^2(x,z) + d^2(y,p) + d^2(x,p) + d^2(y,z)$$

for all $x, y, z, p \in X$. Specifically, the following is Theorem 6 of [26]:

THEOREM 10.4. *A geodesically connected metric space X is a $CAT(0)$ space if and only if it satisfies the quadrilateral inequality condition.*

CHAPTER 11

ℝ-Trees (Metric Trees)

ℝ-trees are a very special class of CAT(0) spaces. There are many equivalent definitions of ℝ-trees. Here are two of them.

DEFINITION 11.1. An ℝ-*tree* is a metric space X such that for every x and y in X there is a unique arc between x and y and this arc is isometric to an interval in ℝ (i.e., is a geodesic segment).

DEFINITION 11.2. An ℝ-*tree* is a metric space X such that
   (i) there is a unique geodesic segment denoted by $[x, y]$ joining each pair of points x and y in X; and
   (ii) $[y, x] \cap [x, z] = \{x\} \Rightarrow [y, x] \cup [x, z] = [y, z]$.

The following is an immediate consequence of (i) and (ii).
   (iii) If $x, y, z \in X$, there exists a point $w \in X$ such that $[x, y] \cap [x, z] = [x, w]$ (whence by (i), $[x, w] \cap [z, w] = \{w\}$).

Standard examples of ℝ-trees include the "radial" and "river" metrics on $\mathbb{R}^2$. For the radial metric, consider all rays emanating from the origin in $\mathbb{R}^2$. Define the radial distance d_r between $x, y \in \mathbb{R}^2$ as follows:

$$d_r(x, y) = d(x, 0) + d(0, y).$$

(Here d denotes the usual Euclidean distance and 0 denotes the origin.) For the river metric ρ, if two points x, y are on the same vertical line, define $\rho(x, y) = d(x, y)$. Otherwise define $\rho(x, y) = |x_2| + |y_2| + |x_1 - y_1|$, where $x = (x_1, x_2)$ and $y = (y_1, y_2)$.

Much more subtle examples exist; e.g., the real tree of Dress and Terhalle [66].

The concept of an ℝ-tree goes back to a 1977 article of J. Tits [210]. The idea has also been attributed to A. Dress [64], who first studied the concept in 1984 and called it T-theory.

Bestvina in [25] observes that much of the importance of ℝ-trees stems from the fact that in many situations a sequence of negatively curved objects (manifolds, groups) gives rise (in some sense "converges") to an ℝ-tree together with a group acting on it by isometries. There are applications in

© Springer International Publishing Switzerland 2014
W. Kirk, N. Shahzad, *Fixed Point Theory in Distance Spaces*,
DOI 10.1007/978-3-319-10927-5_11

biology and computer science as well. The relationship with biology stems from the construction of phylogenetic trees [195]. Concepts of "string matching" are also closely related with the structure of ℝ-trees [19].

The following theorem yields a characterization of hyperconvex CAT(0) spaces.

THEOREM 11.1 ([116]). *For a metric space X the following are equivalent: (i) X is a complete ℝ-tree; (ii) X is hyperconvex and has unique metric segments.*

It is known that a complete ℝ-tree is a complete CAT(0) space [36, p. 167]. On the other hand, a CAT(0) space has unique metric segments. If it is also hyperconvex, then by Theorem 11.1 it must be a complete ℝ-tree. Thus we have:

THEOREM 11.2. *A CAT(0) space is hyperconvex if and only if it is a complete ℝ-tree.*

A proof that a complete ℝ-tree is injective is given in [137]. Since injective spaces are known to be hyperconvex [12] this also gives (i) $\Rightarrow$ (ii). Another proof that (i) $\Rightarrow$ (ii) is given in Aksoy and Maurizi [6]. Their proof is based on an interesting four point property of metric trees.

DEFINITION 11.3. *A metric space (X,d) is said to satisfy the four point property if for each set of four points $x, y, z, w \in X$ the following holds:*

$$d(x,y) + d(u,w) \le \max\{d(x,u) + d(y,w), d(x,w) + d(y,u)\}.$$

Since one obtains the triangle inequality by taking $u = w$, the four point property is a stronger condition. Dress shows in [64] that a metric space is a complete ℝ-tree if and only if it is complete, connected, and satisfies the four point property.

11.1. The Fixed Point Property for ℝ-Trees

G.S. Young, Jr. proved the following result in 1946. He notes explicitly in [220] that compactness is not needed.

THEOREM 11.3 ([219]). *Let X be an arcwise connected Hausdorff space which is such that every monotone increasing sequence of arcs is contained in an arc. Then X has the fixed point property (for continuous maps).*

In [151], J.C. Mayer and L.G. Oversteegen proved that for a separable metric space (X, d) the following are equivalent:

1. X is an ℝ-tree.
2. X is locally arcwise connected and uniquely arcwise connected metric space.

If a complete ℝ-tree is geodesically bounded, it is easy to see that every monotone increasing sequence of arcs is contained in an arc. In view of this, we have the following.

THEOREM 11.4. *A complete geodesically bounded ℝ-tree has the fixed point property for continuous maps.*

Although the validity of Theorem 11.4 goes back to Young's 1946 result, a more constructive metric approach might be of interest. The following proof is taken from [**120**].

PROOF OF THEOREM 11.4. For $u, v \in X$ let $[u, v]$ denote the (unique) metric segment joining u and v and let $[u, v) = [u, v] \setminus \{v\}$. We associate with each point $x \in X$ a point $\varphi(x)$ as follows. For each $t \in [x, f(x)]$, let $\xi(t)$ be the point of X for which

$$[x, f(x)] \cap [x, f(t)] = [x, \xi(t)].$$

(It follows from the definition of an ℝ-tree that such a point always exists.) If $\xi(f(x)) = f(x)$, take $\varphi(x) = f(x)$. Otherwise it must be the case that $\xi(f(x)) \in [x, f(x))$. Let

$$A = \{t \in [x, f(x)] : \xi(t) \in [x, t]\};$$
$$B = \{t \in [x, f(x)] : \xi(t) \in [t, f(x)]\}.$$

Clearly $A \cup B = [x, f(x)]$. Since ξ is continuous, both A and B are closed. Also $A \neq \emptyset$ as $f(x) \in A$. However the fact that $f(t) \to f(x)$ as $t \to x$ implies $B \neq \emptyset$ (because $t \in A$ implies $d(f(t), f(x)) \geq d(t, x)$). Therefore there exists a point $\varphi(x) \in A \cap B$. If $\varphi(x) = x$, then $f(x) = x$ and we are done. Otherwise $x \neq \varphi(x)$ and

$$[x, f(x)] \cap [x, f(\varphi(x))] = [x, \varphi(x)].$$

Now let $x_0 \in X$, and let $x_n = \varphi^n(x_0)$. Assuming the process does not terminate upon reaching a fixed point of f, by construction the points $\{x_0, x_1, x_2, \cdots\}$ are collinear and thus lie on a subset of X which is isometric with a subset of the real line, i.e., on a geodesic. Since X does not contain a geodesic of infinite length, it must be the case that

$$\sum_{i=0}^{\infty} d(x_i, x_{i+1}) < \infty,$$

and hence $\{x_n\}$ is a Cauchy sequence. Suppose $\lim_{n \to \infty} x_n = x^*$. Then by continuity

$$\lim_{n \to \infty} f(x_n) = f(x^*),$$

and in particular $\{f(x_n)\}$ is a Cauchy sequence. However, by construction,

$$d(f(x_n), f(x_{n+1})) = d(f(x_n), x_{n+1}) + d(x_{n+1}, f(x_{n+1})).$$

Since $\lim_{n \to \infty} d(f(x_n), f(x_{n+1})) = 0$, it follows that $\lim_{n \to \infty} d(f(x_n), x_{n+1}) = d(f(x^*), x^*) = 0$ and $f(x^*) = x^*$.   □

11.2. The Lifšic Character of ℝ-Trees

We now turn to the *Lifšic character* of ℝ-trees. (The present discussion is taken from [**130**].) Balls in X are said to be *c-regular* if the following holds: For each $k < c$ there exist $\mu, \alpha \in (0, 1)$ such that for each $x, y \in X$ and $r > 0$ with $d(x, y) \geq (1 - \mu) r$, there exists $z \in X$ such that

$$(11.1) \qquad B(x; (1 + \mu) r) \bigcap B(y; k(1 + \mu) r) \subset B(z; \alpha r).$$

The *Lifšic character* $\kappa(X)$ of X is defined as follows:

$$\kappa(X) = \sup \{c \geq 1 : \text{balls in } X \text{ are } c\text{-regular}\}.$$

A mapping $f : X \to X$ is said to be *eventually k-lipschitzian* if there exists $n_0 \in \mathbb{N}$ such that $d(f^n(x), f^n(y)) \leq kd(x, y)$ for all $x, y \in X$ and $n \geq n_0$. The Lifšic character is fundamental in metric fixed point theory because of the following result.

THEOREM 11.5 ([**140**]). *Let (X, d) be a complete metric space. Then every eventually k-lipschitzian mapping $f : X \to X$ with $k < \kappa(X)$ has a fixed point if it has a bounded orbit.*

PROOF. (Except for the final paragraph, this is identical to the proof given in [**87**, p. 172].) If $\kappa(X) = 1$, then f^n is a contraction mapping for sufficiently large n and there is nothing to prove. So, suppose $\kappa(X) > 1$. For each $x \in X$, set

$$r(x) = \inf \{r > 0 : B(x; r) \text{ contains an orbit of } f\}.$$

Now let $k < \kappa(X)$, and let $\mu, \alpha \in (0, 1)$ be the numbers associated with k in the definition of k-regular balls. Then given any $x \in X$ there is an integer $m \in \mathbb{N}$ such that

$$d(x, f^m(x)) \geq (1 - \mu) r(x)$$

and there is also a point $y \in X$ such that

$$d(x, f^n(y)) \leq (1 + \mu) r(x), \qquad n = 1, 2, \cdots.$$

Since the balls are k-regular, there exists $z \in X$ such that

$$D := B(x; (1 + \mu) r(x)) \cap B(f^m(x); k(1 + \mu) r(x)) \subseteq B(z; \alpha r(x)).$$

Next observe that for m sufficiently large,

$$d(f^m(x), f^n(y)) \leq kd(x, f^{n-m}(y)) \leq k(1 + \mu) r(x)$$

for all $n > m$. This shows that $\{f^n(y)\}_{n > m}$ is contained in D, and hence in $B(z; \alpha r(x))$. This in turn implies that

$$r(z) \leq \alpha r(x).$$

Also, for any $u \in D$,

$$\begin{aligned} d(z, x) &\leq d(z, u) + d(u, x) \\ &\leq \alpha r(x) + (1 + \mu) r(x) \\ &= A r(x), \end{aligned}$$

where $A = \alpha + 1 + \mu$.

By setting $x = x_0$ and $z = z(x_0)$, it is possible to define a sequence $\{x_n\}$ with $x_{n+1} = z(x_n)$, where $z(x_n)$ is defined via the above procedure. Thus $r(x_n) \leq \alpha^n r(x_0)$ and $d(x_n, x_{n+1}) \leq A r(x_n) \leq \alpha^n r(x_0)$. This proves that $\{x_n\}$ is a Cauchy sequence which has limit, say x^*. Now choose $N \in \mathbb{N}$ so that both f^N and f^{N+1} are lipschitzian. Since $B(x^*; \varepsilon)$ contains an orbit of f for any $\varepsilon > 0$, there exists a sequence $\{y_n\}$ also converging to x^* for which $\lim_{n \to \infty} d(f^N(y_n), f^{N+1}(y_n)) = 0$. It follows that $f^N(x^*) = f^{N+1}(x^*)$; hence $f^N(x^*)$ is a fixed point of f.   □

The Lifšic character is known for many classical Banach spaces. For a Hilbert space it is $\sqrt{2}$. The following is proved in [60].

THEOREM 11.6. *If (X, d) is a complete CAT(0) space, then $\kappa(X) \geq \sqrt{2}$. Moreover, if X is an $\mathbb{R}$-tree, $\kappa(X) = 2$.*

Another proof of the second statement is given in [3, Theorem 3.16]; also a characterization of compact $\mathbb{R}$-trees in terms of metric segments is found there.

In view of Theorem 11.5, if X is a complete bounded CAT(0) space, then every eventually k-lipschitzian mapping $f : X \to X$ with $k < \sqrt{2}$ has a fixed point. The corresponding fact for a complete $\mathbb{R}$-tree is the following.

THEOREM 11.7. *Let X be a complete $\mathbb{R}$-tree and let $f : X \to X$ be eventually uniformly k-lipschitzian for $k < 2$, and assume that f has bounded orbits. Then f has a fixed point.*

For a direct proof of this result (and related facts), see [5]. The significance of the above result lies in the fact that the mapping is not assumed to be continuous. A remarkably stronger result holds if f is assumed to be continuous. (Throughout we use $O(x)$ to denote the orbit of a mapping $f : X \to X$ at a point $x \in X$; thus $O(x) = \{x, f(x), f^2(x), \cdots\}$.)

THEOREM 11.8. *Let (X, d) be a complete $\mathbb{R}$-tree. Suppose $f : X \to X$ is continuous and has bounded orbits, and suppose for all $n \in \mathbb{N}$ sufficiently large,*

$$(11.2) \qquad d(f^n(x), f^n(y)) \leq k_n d(x, y)$$

for all $x, y \in X$, with $\limsup_{n \to \infty} k_n < \infty$. Then some bounded convex subset of X is f-invariant; hence f has a fixed point.

This will be an immediate consequence of Theorem 11.4 and the following result.

THEOREM 11.9 ([130]). *Let (X, d) be an $\mathbb{R}$-tree. Suppose $f : X \to X$ is continuous and has bounded orbits, and suppose for all $n \in \mathbb{N}$ sufficiently large,*

$$(11.3) \qquad d(f^n(x), f^n(y)) \leq k_n d(x, y)$$

for all $x, y \in X$, with $\limsup_{n \to \infty} k_n < \infty$. Then some bounded subtree of X is f-invariant.

PROOF. Fix $x \in X$ and choose $m \in \mathbb{N}$ and $k > 0$ with $\limsup_{n\to\infty} k_n < k$ so that $d(f^n(u), f^n(v)) \leq kd(u,v)$ for all $u, v \in [x, f(x)]$ and $n \geq m$. Let $Y = \bigcup_{i=1}^{\infty} f^i([x, f(x)])$. Since each $f^i([x, f(x)])$ is an arcwise connected subset of X, Y is an arcwise connected subset of X; hence Y itself is an ℝ-tree which is clearly f-invariant. We show that Y is bounded.

Let $\xi(z) = \sup\{d(z, f^n(z)) : n \geq m\}$ for each $z \in [x, f(x)]$. By assumption $\xi(z) < \infty$ for each $z \in [x, f(x)]$. If $z, w \in [x, f(x)]$, then

$$\begin{aligned} d(w, f^n(w)) &\leq d(w, z) + d(z, f^n(z)) + d(f^n(z), f^n(w)) \\ &\leq d(w, z) + \xi(z) + kd(z, w) \end{aligned}$$

for each $n \geq m$. Thus $\xi(w) \leq \xi(z) + (1 + k)d(z, w)$. Reversing the roles of z and w, we conclude

$$|\xi(z) - \xi(w)| \leq (1 + k)d(z, w)$$

for all $z, w \in [x, f(x)]$. Thus ξ is continuous, and since $[x, f(x)]$ is compact,

$$\xi := \sup\{\xi(z) : z \in [x, f(x)]\} < \infty.$$

Now for $1 \leq i < m$, let $\beta_i = \sup\{d(z, f^i(z)) : z \in [x, f(x)]\}$ and let

$$\beta = \max\{\beta_i : i = 1, \cdots, m - 1\}.$$

Since f is continuous, $\beta < \infty$. Also, by construction, given $y \in Y$ there is at least one point $z \in [x, f(x)]$ such that $y \in O(z)$. It follows that $d(z, y) \leq \beta + \xi$. Therefore Y is bounded. Specifically, $Y \subset B(x; \gamma)$, where $\gamma = d(x, f(x)) + \beta + \xi$. $\qquad \square$

Since a nonexpansive mapping satisfies (11.2) for $k_n \equiv 1$, we have the following corollary.

COROLLARY 11.1 (Theorem 4.5 (i) of [73]). *A nonexpansive mapping of a complete ℝ-tree into itself with bounded orbits always has a fixed point.*

REMARK 11.1. *Under the assumptions of Theorem 11.8 it is enough to assume that one orbit of f is bounded. Indeed, the following is true.*

PROPOSITION 11.1. *Let (X, d) be a metric space and suppose $f : X \to X$ has a bounded orbit. Suppose that for all n sufficiently large,*

$$d(f^n(x), f^n(y)) \leq k_n d(x, y)$$

for all $x, y \in X$. Suppose also that $\limsup_{n\to\infty} k_n < \infty$. Then all orbits of f are bounded.

PROOF. Assume there exist $x \in X$ and $r > 0$ such that $O(x) \subset B(x; r)$. Choose $k > 0$ so that $\limsup_{n\to\infty} k_n < k$. Then if $y \in X$ it is possible to choose $m \in \mathbb{N}$ so that for all $n \geq m$,

$$d(f^n(x), f^n(y)) \leq kd(x, y).$$

Then for $n \geq m$,

$$d(x, f^n(y)) \leq d(x, f^n(x)) + d(f^n(x), f^n(y)) \leq r + kd(x, y).$$

This proves that $\{f^n(y)\}_{n\geq m} \subset B(x;\gamma)$, where $\gamma = r + kd(x,y)$. Let

$$\gamma' = \max\{d(x, f^i(y)) : i = 1, \cdots, m-1\}.$$

Then $O(y) \subset B(x;\gamma^*)$, where $\gamma^* = \max\{\gamma, \gamma'\}$. Since y is arbitrary, all orbits of f are bounded. $\qquad\qquad\qquad\qquad\qquad\qquad\qquad\qquad\qquad\qquad\Box$

11.3. Gated Sets

Many of the ideas discussed above, especially those in $\mathbb{R}$-trees, can be couched in a more abstract framework. A subset Y of a metric space X is said to be *gated* [65] if for any point $x \notin Y$ there exists a unique point $x_Y \in Y$ (called the *gate* of x in Y) such that for any $z \in Y$,

$$d(x,z) = d(x,x_Y) + d(x_Y, z).$$

Obviously gated sets in a complete geodesic space are always closed and convex.

It is known [65] that gated subsets of a complete geodesic space X are proximinal nonexpansive retracts of X. Specifically, if A is a gated subset of X, then the mapping that associates with each point x in X its gate in A (i.e., the gate-map, or "nearest point map") is nonexpansive. Several other properties of gated sets can be found, for example, in [212, p. 98]. In particular it can be easily shown by induction that the family of gated sets in a complete geodesic space X has the *Helly property*. Thus if $S_1, \cdots, S_n$ is a collection of pairwise intersecting gated sets in X, then $\cap_{i=1}^{n} S_i \neq \emptyset$.

The gated subsets of an $\mathbb{R}$-tree are precisely its closed and convex subsets. Thus the following results apply to $\mathbb{R}$-trees.

PROPOSITION 11.2 ([73]). *Let (X,d) be a complete geodesic space, and let $\{H_\alpha\}_{\alpha\in\Lambda}$ be a collection of nonempty gated subsets of X which is directed downward by set inclusion. If X (or more generally, some H_α) does not contain a geodesic ray, then $\cap_{\alpha\in\Lambda} H_\alpha \neq \emptyset$.*

PROPOSITION 11.3 ([73]). *Let (X,d) be a complete geodesic space, and let $\{H_n\}$ be a descending sequence nonempty gated subsets of X. If $\{H_n\}$ has a bounded selection, then $\cap_{n=1}^{\infty} H_n \neq \emptyset$.*

The following is given in [146].

THEOREM 11.10. *Let (X,d) be a complete $\mathbb{R}$-tree and K a closed convex subset of X. Then $I_K(x)$ is a closed convex set for each $x \in K$, where $I_K(x)$ is the metrically inward set of K at x defined by*

$$I_K(x) = \{z \in X : z = x \text{ or } \exists\, y \in X,\ y \neq x \text{ such that } d(x,z) = d(x,y) + d(y,z)\}.$$

REMARK 11.2. *In a complete $\mathbb{R}$-tree X, if K is a gated subset of X, then $I_K(x)$ is also gated for each $x \in K$.*

11.4. Best Approximation in ℝ-Trees

Ky Fan's classical best approximation theorem [81] asserts that if K is a nonempty compact convex subset of a normed linear space X and if $f : K \to X$ is continuous, then there exists a point $x^* \in K$ such that

$$\|x^* - f(x^*)\| = \inf\{\|x - f(x^*)\| : x \in K\}.$$

Over the years this theorem has been extended in various ways. See, e.g., Singh et al. [204] for a discussion.

There have been two recent approaches to best approximation for set-valued mappings in ℝ-trees. In [123] Fan's best approximation theorem is extended to upper semicontinuous mappings in an ℝ-tree. The proof given in [123] is constructive—a modification of the proof of Theorem 11.4—although as we note below there is a nice topological approach. A second approach is found in [145], where it is shown that a lower semicontinuity assumption also suffices.

We begin with the approach of [123]. Once again we assume that the space X is geodesically bounded, that is, we assume that X does not contain a geodesic of infinite length.

THEOREM 11.11. *Suppose X is a closed convex subset of a complete ℝ-tree Y, and suppose X is geodesically bounded. Let $T : X \to 2^Y$ be an upper semicontinuous mapping whose values are nonempty closed convex subsets of X. Then there exists a point $x^* \in X$ such that*

$$dist(x^*, T(x^*)) = \inf_{x \in X} dist(x, T(x^*)).$$

For a subset B of a metric space Y, $N_\varepsilon(B) = \{x \in Y : dist(x, B) \le \varepsilon\}$. We will need the following lemma.

LEMMA 11.1. *Under the assumptions of Theorem 11.11, let f be the nearest point selection of T. Then if $t_n \to t$, either $f(t_{n_i}) \to f(t)$ for some subsequence $\{t_{n_i}\}$ of $\{t_n\}$, or for n sufficiently large, $f(t) \in [t_n, f(t_n)]$.*

PROOF. Suppose $\{f(t_n)\}$ is bounded away from $f(t)$, say $d(f(t_n), f(t)) \ge \varepsilon > 0$ for all n. By upper semicontinuity of T there exists $\rho > 0$ such that

$$d(t_n, t) < \rho \Rightarrow T(t_n) \subset N_\varepsilon(T(t)).$$

Hence there exists a point $u_n \in T(t)$ such that $d(f(t_n), u_n) < \varepsilon$. Since $f(t) \in [t, u_n]$, it follows that $f(t) \in [t, f(t_n)]$. Therefore for n sufficiently large any segment joining t_n to a point of $T(t)$ must pass through $f(t)$, whence $f(t) \in [t_n, f(t_n)]$. □

PROOF OF THEOREM 11.11. For $u, v \in X$ we let $[u, v]$ denote the (unique) metric segment joining u and v and let $[u, v) = [u, v] \setminus \{v\}$. For

each $x \in X$, let $f(x)$ be the nearest point selection of T. Thus $f(x) \in T(x)$ and
$$d(x, f(x)) = dist(x, T(x)).$$
We associate with each point $x \in X$ a point $\varphi(x)$ as follows. For each $t \in [x, f(x)] \cap X$, let $\xi(t)$ be the point of X for which
$$[x, f(x)] \cap [x, f(t)] = [x, \xi(t)].$$
(It follows from the definition of an ℝ-tree that such a point always exists.) If $\xi(f(x)) = f(x)$, take $\varphi(x) = f(x)$. Otherwise it must be the case that $\xi(f(x)) \in [x, f(x))$. Let
$$A = \{t \in [x, f(x)] \cap X : \xi(t) \in [x, t]\};$$
$$B = \{t \in [x, f(x)] \cap X : \xi(t) \in [t, f(x)]\}.$$
Clearly $A \cup B = [x, f(x)] \cap X$.

Now let $t \in [x, f(x)] \cap X$ and let $\varepsilon > 0$. Choose $\rho > 0$ so that
$$d(t, t') < \rho \Rightarrow T(t') \subset N_\varepsilon(T(t)).$$
Then it is easy to see that either $d(f(t), f(t')) < \varepsilon$ or $f(t) \in [t', f(t')]$.

One can use the above fact to show that both A and B are closed. Also $A \neq \emptyset$ as $f(x) \in A$. The fact that $B \neq \emptyset$ also follows from the above upon letting $t \to x$. Therefore there exists a point $\varphi(x) \in A \cap B$. If $\varphi(x) = x$, then $f(x) = x$ and we are done. Otherwise $x \neq \varphi(x)$ and
$$[x, f(x)] \cap [x, f(\varphi(x))] = [x, \varphi(x)].$$

Now let $x_0 \in X$, and let $x_n = \varphi^n(x_0)$. If the process terminates, then either one has reached a fixed point of T, or one has reached a point x^* for which $[x^*, f(x^*)] \cap X = \{x^*\}$. In the latter case, clearly
$$dist(x^*, T(x^*)) = \inf_{x \in X} dist(x, T(x^*)).$$
So we assume the process does not terminate. The points $\{x_0, x_1, x_2, \cdots\}$ have been constructed so that they lie on a geodesic. Since X does not contain a geodesic of infinite length, it must be the case that
$$\sum_{i=0}^{\infty} d(x_i, x_{i+1}) < \infty,$$
and hence that $\{x_n\}$ is a Cauchy sequence. Suppose $\lim_{n \to \infty} x_n = x^*$. By construction

(11.4) $d(f(x_n), x_{n+1}) + d(x_{n+1}, x^*) + d(x^*, f(x^*)) = d(f(x_n), f(x^*))$.

We now invoke Lemma 11.1. Clearly (11.4) precludes the possibility that $f(x^*) \in [x_n, f(x_n)]$ for n sufficiently large. On the other hand, if $\lim_{i \to \infty} f(x_{n_i}) = f(x^*)$ for some subsequence $\{x_{n_i}\}$ of $\{x_n\}$, then (11.4) implies $d(x^*, f(x^*)) = 0$, whence $x^* \in T(x^*)$. □

As we remarked above, there is a nice topological approach to Theorem 11.11. Let X be a connected Hausdorff space. A point $p \in X$ *separates* $u, v \in X$ if u and v are contained in disjoint open subsets of $X \setminus \{p\}$. If $e \in X$ it is possible to define a relation Γ_e on $X \times X$ in the following way. (Here $\Delta(X \times X)$ denotes the diagonal in $X \times X$.)

$$\Gamma_e = (\{e \times X\}) \cup \Delta(X \times X) \cup \{(x, y) : x \text{ separates } e \text{ from } y\}.$$

It is known [214] that Γ_e is a partial order.

A connected Hausdorff space X is said to satisfy *property* D(3) if the following condition holds: If A and B are disjoint closed connected subsets of X, then there exists $z \in X$ such that z separates A and B.

Over 50 years ago L.E. Ward, Jr. proved the following result.

THEOREM 11.12 ([214]). *Suppose X is a connected Hausdorff space that satisfies property $D(3)$. Suppose also that there exists $e \in X$ such that, relative to Γ_e, each chain in X has a maximal element and a minimal element. Let $T : X \to 2^X$ be an upper semicontinuous mapping whose values are nonempty closed connected subsets of X. Then T has a fixed point.*

As we show below, Theorem 11.11 is an easy consequence of Ward's theorem.

ANOTHER PROOF OF THEOREM 11.11 ([123]). Since an ℝ-tree is a CAT(0) space, the nearest point map P of Y onto X is nonexpansive by Proposition 9.5. Hence the map $P \circ T : X \to 2^X$ is upper semicontinuous and has a fixed point x^* by Theorem 11.12. Thus there exists $y \in T(x^*)$ such that $P(y) = x^*$. However since P is the nearest point map, it must be the case that $P(y) = x^*$ for all $y \in T(x^*)$. If $x^* \in T(x^*)$ we are finished. Otherwise, choose $y_1 \in T(x^*)$ such that $d(x^*, y_1) = dist(x^*, T(x^*))$. Then if $x \in X$ and $x \neq x^*$,

$$dist(x^*, T(x^*)) = d(x^*, y_1) < d(x, x^*) + d(x^*, y_1) = dist(x, T(x^*)).$$

$\square$

In fact, the following extension of Theorem 11.12 actually gives a topological version of Fan's best approximation theorem. In this theorem

$$x\Gamma_e := \{z \in X : x \leq z\}$$

where $x \leq z$ means $(x, z) \in \Gamma_e$.

THEOREM 11.13 ([123]). *Suppose Y is a connected Hausdorff space that satisfies property $D(3)$ and suppose X is a closed and connected subset of Y. Suppose also that there exists $e \in X$ such that, relative to Γ_e, each chain in X has a maximal element and a minimal element. Let $T : X \to 2^Y$ be an* *upper semicontinuous mapping whose values are nonempty closed connected subsets of Y. Then either T has a fixed point, or there exists $x \in \partial X$ such that $T(x) \subset x\Gamma_e \setminus \{x\}$.*

The following KKM principle for trees is also proved in [123]. It can also be used to give yet another proof of Theorem 11.11. In this theorem, $conv_Y(F)$ denotes the intersection of all closed convex subsets of Y that contain F.

THEOREM 11.14 ([123]). *Suppose X is a closed convex subset of a complete $\mathbb{R}$-tree Y, and suppose $G : X \to 2^Y$ has nonempty closed values. Suppose also that for each finite $F \subset X$,*

$$conv_Y(F) \subset \bigcup_{x \in F} G(x).$$

Then $\{G(x)\}_{x \in X}$ has the finite intersection property. Moreover, if X is geodesically bounded, $\bigcap_{x \in X} G(x) \neq \emptyset$.

We now turn to the results of Markin [145]. Let X be a topological space, Y a metric space, and $T : X \to 2^Y$ a mapping with nonempty values. T is said to be *almost lower semicontinuous* if given $\varepsilon > 0$, for each $x \in X$ there is a neighborhood $U(x)$ of x such that $\bigcap_{y \in U(x)} N_\varepsilon(T(y)) \neq \emptyset$. It is easy to check that a mapping which is lower semicontinuous in the usual sense is also almost lower semicontinuous.

THEOREM 11.15 ([145]). *Suppose X is a closed convex subset of a complete $\mathbb{R}$-tree Y, and suppose X is geodesically bounded. Let $T : X \to 2^Y$ be an almost lower semicontinuous mapping whose values are nonempty bounded closed convex subsets of Y. Then there exists a point $x^* \in X$ such that*

$$dist(x^*, T(x^*)) = \inf_{x \in X} dist(x, T(x^*)).$$

The proof of Theorem 11.15 is based on Proposition 11.3 and the following selection theorem for $\mathbb{R}$-trees.

THEOREM 11.16 ([145]). *Let X be a paracompact topological space, Y a complete $\mathbb{R}$-tree, and $T : X \to 2^Y$ an almost lower semicontinuous mapping whose values are nonempty bounded closed convex subsets of Y. Then T has a continuous selection.*

11.5. Applications to Graph Theory

A *graph* is an ordered pair (V, E) where V is a set and E is a binary relation on V $(E \subseteq V \times V)$. Elements of E are called *edges*. We are concerned here with (undirected) graphs that have a "loop" at every vertex (i.e., $(a, a) \in E$ for each $a \in V$) and no "multiple" edges. Such graphs are called *reflexive*. In this case $E \subseteq V \times V$ corresponds to a reflexive (and symmetric) binary relation on V.

For a graph $G = (V, E)$ a map $f : V \to V$ is *edge-preserving* if $(a, b) \in E \Rightarrow (f(a), f(b)) \in E$. For such a mapping we simply write $f : G \to G$. There is a standard way of *metrizing* connected graphs; let each edge have length one and take distance $d(a, b)$ between two vertices a and b to be the length of the shortest path joining them. With this metric edge-preserving mappings become precisely the *nonexpansive* mappings. (Keep in mind that in a reflexive graph an edge-preserving map may collapse edges between distinct points since loops are allowed.)

We now turn to the classical Fixed Edge Theorem and show how it is a consequence of Theorem 11.4.

THEOREM 11.17 ([**161**]). *Let G be a reflexive graph that is connected, contains no cycles, and contains no infinite paths. Then every edge-preserving map of G into itself fixes an edge.*

PROOF ([**73**]). Suppose $f : G \to G$ is edge-preserving. Since a connected graph with no cycles is a tree, one can construct from the graph G an ℝ-tree $\mathfrak{X}$ by identifying each (nontrivial) edge with a unit interval of the real line and assigning the shortest path distance to any two points of $\mathfrak{X}$. It is easy to see that with this metric $\mathfrak{X}$ is complete. One can now extend f affinely on each edge to the corresponding unit interval, and the resulting mapping $\bar{f}$ is a nonexpansive (hence continuous) mapping of $\mathfrak{X} \to \mathfrak{X}$. Thus $\bar{f}$ has a fixed point z by Theorem 11.4. Moreover, since $\mathfrak{X}$ has unique metric segments and $\bar{f}$ is nonexpansive, the fixed point set F of $\bar{f}$ is convex (and closed). It follows from this that either F contains a vertex of G, or z is the midpoint of a unit interval of $\mathfrak{X}$ in which case f must leave the corresponding edge fixed.   □

An application of Baillon's theorem [**17**] about commuting families of nonexpansive mappings in hyperconvex metric space tells us even more. For details, see [**73**].

THEOREM 11.18 ([**73**]). *Let G be a reflexive graph that is connected, contains no cycles, and contains no infinite paths. Suppose $\mathfrak{F}$ is a commuting family of edge-preserving mappings of G into itself. Then either:*

(a) *there is a unique edge in G that is left fixed by each member of $\mathfrak{F}$; or*
(b) *some vertex of G is left fixed by each member of $\mathfrak{F}$.*

It is likely that the above result is known in a more abstract framework. This seems to be a natural in a metric space context.

Part III

Beyond Metric Spaces

b-Metric Spaces

12.1. Introduction

In 1993 another axiom for semimetric spaces, which is weaker than the triangle inequality, was put forth by Czerwik [**58**] with a view of generalizing the Banach contraction mapping theorem. This same relaxation of the triangle inequality is also discussed in Fagin et al. [**79**], who call this new distance measure *nonlinear elastic matching* (NEM). The authors of that paper remark that this measure has been used, for example, in [**55**] for trademark shapes and in [**152**] to measure ice floes. Later Q. Xia [**218**] used this semimetric distance to study the optimal transport path between probability measures. Xia has chosen to call these spaces quasimetric spaces, which is the term used in the book by Heinonen [**92**].

DEFINITION 12.1 ([**18, 58**]). A semimetric space (X, d) is said to be a *b-metric space* (or *quasimetric space*) if there exists $s \geq 1$ such that for each $x, y, z \in X$,

$$(12.1) \qquad d(x, y) \leq s\left[d(x, z) + d(z, y)\right].$$

Obviously a b-metric space for $s = 1$ is precisely a metric space. We note also that these spaces are called s-relaxed$_t$ metric spaces in [**78**]. We mention two examples. Other examples are found in the papers cited.

EXAMPLE 12.1 ([**31**]). *Let $p \in (0, 1)$, and let*

$$X = \ell_p(\mathbb{R}) := \left\{ x = \{x_n\} \subset \mathbb{R} : \sum_{n=1}^{\infty} |x_n|^p < \infty \right\}.$$

For $x, y \in X$, set $d(x, y) = \left(\sum_{n=1}^{\infty} |x_n - y_n|^p\right)^{1/p}$. Then (X, d) is a b-metric space with $s = 2^{1/p}$.

The next example follows from the fact that if a and b are positive real numbers and $\beta > 1$, then

$$\left(\frac{a+b}{2}\right)^{\beta} \leq \frac{a^{\beta} + b^{\beta}}{2}$$

and this in turn follows from the fact that the real valued function $x \mapsto x^{\beta}$ is convex.

W. Kirk, N. Shahzad, *Fixed Point Theory in Distance Spaces*,
DOI 10.1007/978-3-319-10927-5_12

EXAMPLE 12.2 ([**218**]). *Suppose (X, d) is a metric space. Let $\beta > 1$, $\lambda \geq 0$, and $\mu > 0$, and for $x, y \in X$ define $J(x, y) = \lambda d(x, y) + \mu d(x, y)^{\beta}$. Typically J is not a metric on X. However (X, J) is a b-metric space with $s = 2^{\beta - 1}$. Indeed, for any $z \in X$,*

$$
\begin{aligned}
J(x, y) &= \lambda d(x, y) + \mu d(x, y)^{\beta} \\
&\leq \lambda [d(x, z) + d(z, y)] + \mu [d(x, z) + d(z, y)]^{\beta} \\
&\leq \lambda [d(x, z) + d(z, y)] + 2^{\beta - 1} \mu \left[d(x, z)^{\beta} + d(z, y)^{\beta} \right] \\
&\leq 2^{\beta - 1} [J(x, z) + J(z, y)].
\end{aligned}
$$

DEFINITION 12.2 ([**78**]). A semimetric space (X, d) is said to have the *metric boundedness property* if there exists a metric ρ on X and positive constants s_1 and s_2 such that for each $x, y \in X$,

$$
s_1 \rho(x, y) \leq d(x, y) \leq s_2 \rho(x, y).
$$

By adjusting the metric ρ to $s_1^{-1} \rho$ it is clear that we may assume $s_1 = 1$, in which case d is said to be s_2-*metric bounded*.

It is immediate that the metric boundedness property implies that the semimetric is a *b*-metric, since in this case for each $x, y, z \in X$,

$$
\begin{aligned}
d(x, y) &\leq s_2 \rho(x, y) \\
&\leq s_2 [\rho(x, z) + \rho(z, y)] \\
&\leq s_2 [d(x, z) + d(z, y)].
\end{aligned}
$$

It is also noted by Fagin et al. in [**78**] that while the converse is not true, rather surprisingly the converse is true if one replaces the relaxed triangle inequality (12.1) in the definition of a *b*-metric with the *relaxed polygonal inequality*, which asserts that there is a constant $s \geq 1$ such that for all $n \in \mathbb{N}$ and $x, y, x_1, \cdots, x_{n-1} \in X$,

$$
d(x, y) \leq s[d(x, x_1) + d(x_1, x_2) + \cdots + d(x_{n-1}, y)].
$$

This leads to the following definition.

DEFINITION 12.3 ([**78**]). An s-*relaxed$_p$ metric* is a semimetric space (X, d) for which d satisfies the relaxed polygonal inequality, that is,

$$
d(x, y) \leq s[d(x, x_1) + d(x_1, x_2) + \cdots + d(x_{n-1}, y)]
$$

for all $x, x_1, \cdots, x_{n-1}, y \in X$.

We discuss these facts further in Sect. 12.8.

Since every *b*-metric space is a semimetric space, we adopt the terminology and notation of Chap. 1. Also it is easy to see that any *b*-metric space satisfies Wilson's Axiom V by simply defining r_k to be $\dfrac{k}{s}$ and thus, according to Wilson, they are metrizable. See [**78**] for a further discussion of metrizability of *b*-metric spaces.

12.2. Banach's Theorem in a b-Metric Space

THEOREM 12.1. *Let (X, d) be a complete semimetric space which satisfies Wilson's Axiom* III *(see Chap. 1), suppose $k \in (0,1)$, and suppose $f : X \to X$ satisfies*

$$d(f(x), f(y)) \le kd(x, y)$$

for all $x, y \in X$. Suppose some orbit $O(x) := \{x, f(x), f^2(x), \cdots\}$ is bounded. Then f has a unique fixed point $x^* \in X$, and $\lim_{n \to \infty} f^n(u) = x^*$ for each $u \in X$.

PROOF. Let $\varepsilon > 0$ and let $M = diam(O(x))$. Choose $m \in \mathbb{N}$ so that $k^i \le \varepsilon/M$ for $i \ge m$. Then if $j > i \ge m$,

$$d(f^i(x), f^j(x)) \le k^m d(f^{i-m}(x), f^{j-m}(x)) \le k^m M \le \varepsilon.$$

This proves that $\{f^n(x)\}$ is a Cauchy sequence. Therefore there exists $x^* \in X$ such that $\lim_{n \to \infty} f^n(x) = x^*$, and by Proposition 1.1, x^* is unique. Since f is continuous, it follows that $f(x^*) = x^*$. Moreover if $u \in X$, $d(f^n(u), x^*) = d(f^n(u), f^n(x^*)) \le k^n d(u, x^*) \to 0$, so $\lim_{n \to \infty} f^n(u) = x^*$. $\qquad\square$

In a b-metric space the assumption that $O(x)$ is bounded may be dropped. As we noted above a b-metric space satisfies Wilson's Axiom III. Indeed, the following is essentially Theorem 1 of [58].

THEOREM 12.2. *Let (X, d) be a complete b-metric space with constant $s > 1$, and suppose $f : X \to X$ satisfies*

$$d(f(x), f(y)) \le \varphi(d(x, y))$$

for each $x, y \in X$, where $\varphi : [0, \infty) \to [0, \infty)$ is increasing and satisfies

$$\lim_{n \to \infty} \varphi^n(t) = 0$$

for each $t > 0$. Then f has a unique fixed point $x^* \in X$, and $\lim_{n \to \infty} f^n(x) = x^*$ for each $x \in X$.

PROOF. (cf., [58]) First we observe that the assumptions on φ imply that

$$\lim_{t \to 0^+} \varphi(t) = 0,$$

so f is continuous. Now let $x \in X$ and let $\varepsilon > 0$ be arbitrary. Choose $n \in \mathbb{N}$ so that $\varphi^n(\varepsilon) < \varepsilon/2s$. Put $g = f^n$ and for each $m \in \mathbb{N}$ set $x_m = g^m(x)$. Then

$$d(x_{m+1}, x_m) = d(g^m(gx), g^m(x)) \le \varphi^{nm}(d(g(x), x))$$

so $\lim_{m \to \infty} d(x_{m+1}, x_m) = 0$.

Now choose $m \in \mathbb{N}$ so that $d(x_{m+1}, x_m) < \varepsilon/2s$ and let $u \in B(x_m; \varepsilon)$. Then

$$d(g(u), g(x_m)) \le \varphi^n(d(u, x_m)) \le \varphi^n(\varepsilon) < \varepsilon/2s$$

and
$$d\left(g\left(x_m\right), x_m\right) = d\left(x_{m+1}, x_m\right) < \varepsilon/2s.$$
By the relaxed triangle inequality,
$$\begin{aligned} d\left(g\left(u\right), x_m\right) &\leq s\left[d\left(g\left(u\right), g\left(x_m\right)\right) + d\left(g\left(x_m\right), x_m\right)\right] \\ &< s\left[\frac{\varepsilon}{2s} + \frac{\varepsilon}{2s}\right] = \varepsilon. \end{aligned}$$
Therefore $g : B\left(x_m; \varepsilon\right) \to B\left(x_m; \varepsilon\right)$. It follows that if $j, t \geq m$,
$$\begin{aligned} d\left(x_t, x_j\right) &\leq s\left[d\left(x_t, x_m\right) + d\left(x_m, x_j\right)\right] \\ &\leq 2s\varepsilon. \end{aligned}$$

This proves that $\{x_m\}$ is a Cauchy sequence, so there exists $x^* \in X$ such that $\lim_{m \to \infty} x_m = x^*$. Also, continuity of f implies continuity of g, so
$$x^* = \lim_{m \to \infty} x_m = \lim_{x \to \infty} x_{m+1} = \lim_{m \to \infty} g\left(x_m\right) = g\left(x^*\right).$$
Since
$$d\left(g\left(x^*\right), g\left(y^*\right)\right) \leq \varphi^n\left(d\left(x^*, y^*\right)\right) < d\left(x^*, y^*\right)$$
if $x^* \neq y^*$, it is clear that g has exactly one fixed point. Also, since
$$d\left(x^*, g^m\left(x\right)\right) = d\left(g^m\left(x^*\right), g^m\left(x\right)\right) \leq \varphi^{nm}\left(d\left(x^*, x\right)\right) \to 0 \text{ as } m \to \infty,$$
$\{g^m\left(x\right)\}$ converges to x^* for all $x \in X$. However, by continuity of f,
$$f\left(x^*\right) = \lim_{m \to \infty} f\left(x_m\right) = \lim_{m \to \infty} f\left(g^m\left(x\right)\right) = \lim_{m \to \infty} g^m\left(f\left(x\right)\right) = x^*.$$
Therefore x^* is also the unique fixed point of f. Finally, since for any $x \in X$ and $r \in \{0, 1, \cdots, n-1\}$,
$$f^{nm+r}\left(x\right) = g^m\left(f^r\left(x\right)\right) \to x^* \text{ as } m \to \infty,$$
it follows that $\lim_{m \to \infty} f^m\left(x\right) = x^*$. $\square$

Remark. Theorem 12.2 reveals the extent to which the Banach contraction mapping theorem does NOT depend on the triangle inequality.

12.3. b-Metric Spaces Endowed with a Graph

The material in this section is taken from [**193**], motivated in turn by ideas introduced by Jachymski in [**100**]. We refer to [**29**, **193**] for further discussion and citations.

Throughout (X, d) denotes a b-metric space with coefficient $s \geq 1$ and Δ is the diagonal of the cartesian product $X \times X$. G is a directed graph such that the set $V(G)$ of its vertices coincides with X, and the set $E(G)$ of its edges contains all loops, i.e., $E(G) \supseteq \Delta$. Assume that G has no parallel edges (i.e., multiple edges). We assign to each edge having vertices x and y a unique element $d(x, y)$.

We will also use the following concept introduced by Matkowski [**148**, p. 68] in his well-known generalization of Banach's contraction mapping principle.

Let $\varphi : \mathbb{R}^+ \to \mathbb{R}^+$. Consider the following properties:

$(i)_\varphi$ $t_1 \leq t_2 \implies \varphi(t_1) \leq \varphi(t_2)$, $\forall t_1, t_2 \in \mathbb{R}^+$;

$(ii)_\varphi$ $\varphi(t) < t$ for $t > 0$;

$(iii)_\varphi$ $\varphi(0) = 0$;

$(iv)_\varphi$ $\lim_{n\to\infty} \varphi^n(t) = 0$ for all $t \geq 0$;

$(v)_\varphi$ $\sum_{n=0}^{\infty} \varphi^n(t)$ converges for all $t > 0$.

It is easily seen that: $(i)_\varphi$ and $(iv)_\varphi$ imply $(ii)_\varphi$; $(i)_\varphi$ and $(ii)_\varphi$ imply $(iii)_\varphi$.

We recall that a function φ satisfying $(i)_\varphi$ and $(iv)_\varphi$ is said to be a *comparison function*. A function φ satisfying $(i)_\varphi$ and $(v)_\varphi$ is known as (c)-*comparison function*. Any (c)-comparison function is a comparison function but converse may not be true. For example, $\varphi(t) = \frac{t}{1+t}; t \in \mathbb{R}^+$ is a comparison function but not a (c)-comparison function. On the other hand, define $\varphi(t) = \frac{t}{2}; 0 \leq t \leq 1$ and $\varphi(t) = t - \frac{1}{2}; t > 1$, then φ is a (c)-comparison function. For details on φ contractions we refer the readers to [**23, 191**].

Berinde [**24**] took further step to investigate φ contractions when the framework was taken to be a b-metric space and for some technical reasons he had to introduce the notion of b-comparison function in particular he obtained some estimations for rate of convergence [**24**].

Now, we introduce the following definition.

DEFINITION 12.4. *We say that a mapping $f : X \to X$ is a b-(φ, G) contraction if for all $x, y \in X$:*

(12.2) $\qquad (f(x), f(y)) \in E(G)$ *whenever* $(x, y) \in E(G)$;

(12.3) $\qquad d(f(x), f(y)) \leq \varphi(d(x, y))$ *whenever* $(x, y) \in E(G)$,

where $\varphi : \mathbb{R}^+ \to \mathbb{R}^+$ is a comparison function.

REMARK 12.1. *A mapping $f : X \to X$ is called a Banach G-contraction if (i) $\forall x, y \in X((x, y) \in E(G) \Rightarrow (f(x), f(y)) \in E(G))$, (ii) $\exists k \in (0, 1)$ such that $\forall x, y \in X$, $(x, y) \in E(G) \Rightarrow d(f(x), f(y)) \leq kd(x, y)$. Note that a Banach G-contraction is a b-(φ, G) contraction.*

EXAMPLE 12.3. *Any constant mapping $f : X \to X$ is a b-(φ, G) contraction for any graph G with $V(G) = X$.*

EXAMPLE 12.4. *Any self-mapping f on X is trivially a b-(φ, G_1) contraction, where $G_1 = (V(G), E(G)) = (X, \Delta)$.*

EXAMPLE 12.5. *Let $X = \mathbb{R}$ and define $d : X \times X \to \mathbb{R}$ by $d(x, y) = |x - y|^2$. Then d is a b-metric on X with $s = 2$. Further, set $f(x) = \frac{x}{2}$, for all $x \in X$. Then f is a b-(φ, G_0) contraction with $\varphi(t) = t/4$ and $G_0 = (X, X \times X)$. Note that d is not a metric on X.*

DEFINITION 12.5. *Sequences $\{x_n\}$ and $\{y_n\}$ in X are said to be equivalent sequences if $\lim_{n\to\infty} d(x_n, y_n) = 0$, and if each of them is a Cauchy sequence then they are called Cauchy equivalent.*

As an immediate consequence of Definition 12.5, we get the following fact.

REMARK 12.2. *Let $\{x_n\}$ and $\{y_n\}$ be equivalent sequences in X. (i) If $\{x_n\}$ converges to x, then $\{y_n\}$ also converges to x and vice versa. (ii) $\{y_n\}$ is a Cauchy sequence whenever $\{x_n\}$ is a Cauchy sequence and vice versa.*

Now we recollect some preliminaries from graph theory which we need for the sequel. Let $G = (V(G), E(G))$ be a directed graph. By letter $\tilde{G}$ we denote the undirected graph obtained from G by ignoring the direction of edges. If x and y are vertices in a graph G, then a path in G from x to y of length l is a sequence $\{x_i\}_{i=0}^{l}$ of $l+1$ vertices such that $x_0 = x, x_l = y$ and $(x_{i-1}, x_i) \in E(G)$ for $i = 1, \cdots, l$. A graph G is called *connected* if there is a path between any two vertices. G is weakly connected if $\tilde{G}$ is connected. For a graph G such that $E(G)$ is symmetric and x is a vertex in G, the subgraph G_x consisting of all edges and vertices which are contained in some path beginning at x is known as a component of G containing x. So that $V(G_x) = [x]_{\tilde{G}}$, where $[x]_{\tilde{G}}$ is the equivalence class of a relation R defined on $V(G)$ by the rule: yRz if there is a path in G from y to z. Clearly, G_x is connected. A graph G is known as (C)-graph in X [7] if for any sequence $\{x_n\}$ in X with $x_n \longrightarrow x$ and $(x_n, x_{n+1}) \in E(G)$ for $n \in \mathbb{N}$ then there exists a subsequence $\{x_{n_j}\}$ of $\{x_n\}$ such that $(x_{n_j}, x) \in E(G)$ for $j \in \mathbb{N}$.

PROPOSITION 12.1. *Let $f : X \to X$ be a b-(φ, G) contraction, where $\varphi : \mathbb{R}^+ \to \mathbb{R}^+$ is a comparison function. Then:*

(i) *f is a b-$\left(\varphi, \tilde{G}\right)$ contraction and also a b-(φ, G^{-1}) contraction;*

(ii) *$[x_0]_{\tilde{G}}$ is f-invariant, and $f \mid_{[x_0]_{\tilde{G}}}$ is a b-$\left(\varphi, \tilde{G}_{x_0}\right)$ contraction provided $x_0 \in X$ is such that $f(x_0) \in [x_0]_{\tilde{G}}$.*

PROOF. (i) This is immediate from the symmetry of a b-metric

(ii) Let $x \in [x_0]_{\tilde{G}}$. Then there is a path $x = z_0, z_1, \cdots, z_l = x_0$ from x to x_0. Since f is a b-(φ, G) contraction, $(f(z_{i-1}), f(z_i)) \in E(G)$ for all $i = 1, 2, \cdots, l$. Thus $f(x) \in [f(x_0)]_{\tilde{G}} = [x_0]_{\tilde{G}}$. Suppose $(x, y) \in E\left(\tilde{G}_{x_0}\right)$. Then again since f is a b-(φ, G) contraction, $(f(x), f(y)) \in E(G)$. But $[x_0]_{\tilde{G}}$ is f-invariant, so we conclude that $(f(x), f(y)) \in E\left(\tilde{G}_{x_0}\right)$. Condition (12.3) is satisfied automatically as $\tilde{G}_{x_0}$ is a subgraph of G.

$\square$

Henceforth we assume that the coefficient of the b-comparison function is at least as large as the coefficient s of the b-metric.

LEMMA 12.1. *Let $f : X \to X$ be a b-(φ, G) contraction, where $\varphi : \mathbb{R}^+ \to \mathbb{R}^+$ is a comparison function. Then given any $x \in X$ and $y \in [x]_{\tilde{G}}$, the sequences $\{f^n(x)\}$ and $\{f^n(y)\}$ are equivalent.*

PROOF. Assume $x \in X$ and $y \in [x]_{\tilde{G}}$. Then there exists a path $\{x_i\}_{i=0}^l$ in $\tilde{G}$ from x to y with $x_0 = x$, $x_l = y_0$ and $(x_{i-1}, x_i) \in E\left(\tilde{G}\right)$. From Proposition 12.1 f is a b-$\left(\varphi, \tilde{G}\right)$ contraction. Therefore

$$(f^n(x_{i-1}), f^n(x_i)) \in E\left(\tilde{G}\right) \Rightarrow$$
$$d(f^n(x_{i-1}), f^n(x_i)) \leq \varphi\left(d\left(f^{n-1}(x_{i-1}), f^{n-1}(x_i)\right)\right)$$

for all $n \in \mathbb{N}$ and $i = 1, 2, \cdots, l$. Hence

$$(12.4) \qquad d(f^n(x_{i-1}), f^n(x_i)) \leq \varphi^n(d(x_{i-1}, x_i))$$

for all $n \in \mathbb{N}$ and $i = 1, 2, \cdots, l$. We observe that $\{f^n(x_i)\}_{i=0}^l$ is a path in $\tilde{G}$ from $f^n(x)$ to $f^n(y)$. Using (12.1) and (12.4) we have

$$d(f^n(x), f^n(y)) \leq \sum_{i=1}^l s^i d(f^n(x_{i-1}), f^n(x_i))$$
$$\leq \sum_{i=1}^l s^i \varphi^n(d(x_{i-1}, x_i)).$$

Letting $n \to \infty$, we obtain $d(f^n(x), f^n(y)) \to 0$. $\qquad \square$

PROPOSITION 12.2. *Let $f : X \to X$ be a b-(φ, G) contraction, where $\varphi : \mathbb{R}^+ \to \mathbb{R}^+$ is a comparison function. Suppose $f(z_0) \in [z_0]_{\tilde{G}}$ for some $z_0 \in X$. Then $\{f^n(z_0)\}$ is a Cauchy sequence.*

PROOF. Since $f(z_0) \in [z_0]_{\tilde{G}}$, let $\{y_i\}_{i=0}^r$ be a path from z_0 to $f(z_0)$. Then following the argument in the previous lemma we arrive at the conclusion

$$d\left(f^n(z_0), f^{n+1}(z_0)\right) \leq \sum_{i=1}^r s^i \varphi^n(d(y_{i-1}, y_i)) \text{ for each } n \in \mathbb{N}.$$

Let $m > n \geq 1$. Then from the above inequality it follows that for $p \geq 1$,

$$(12.5) \quad d\left(f^n(z_0), f^{n+p}(z_0)\right) \leq s d\left(f^n(z_0), f^{n+1}(z_0)\right)$$
$$+ s^2 d\left(f^{n+1}(z_0), f^{n+2}(z_0)\right)$$
$$+ \cdots + s^p d\left(f^{n+p-1}(z_0), f^{n+p}(z_0)\right)$$
$$\leq \frac{1}{s^{n-1}} \left[\sum_{j=n}^{n+p-1} s^j d\left(f^j(z_0), f^{j-1}(z_0)\right) \right]$$
$$\leq \frac{1}{s^{n-1}} \left[\sum_{i=1}^r s^i \sum_{j=n}^{n+p-1} s^j \varphi^j(d(y_{i-1}, y_i)) \right].$$

Denoting for each $i = 1, 2, \cdots, r$

$$S_n^i = \sum_{j=0}^{n} s^j \varphi^j \left(d \left(y_{i-1}, y_i \right) \right), \quad n \geq 1,$$

relation (12.5) becomes

(12.6) $\qquad d \left(f^n \left(z_0 \right), f^{n+p} \left(z_0 \right) \right) \leq \dfrac{1}{s^{n-1}} \left[\sum_{i=1}^{r} \left[S_{n+p-1}^i - S_{n-1}^i \right] \right].$

Since φ is a b-comparison function, for each $i = 1, 2, \cdots, r$,

$$\sum_{j=0}^{\infty} s^j \varphi^j \left(d \left(y_{i-1}, y_i \right) \right) < \infty.$$

Then corresponding to each i, there is a real number S^i such that

$$\lim_{n \to \infty} S_n^i = S^i.$$

In view of this (12.6) gives $d \left(f^n \left(z_0 \right), f^{n+p} \left(z_0 \right) \right) \to 0$ as $n \to \infty$. This proves that $\{ f^n \left(z_0 \right) \}$ is a Cauchy sequence in X. $\qquad \square$

DEFINITION 12.6 (cf., [192]). Let $f : X \to X$, let $y \in X$ and suppose the sequence $\{ f^n \left(y \right) \}$ in X is such that $f^n \left(y \right) \to x^*$ with $\left(f^n \left(y \right), f^{n+1} \left(y \right) \right) \in E(G)$ for $n \in \mathbb{N}$. We say that a graph G is (C_f)-*graph* if there exists a subsequence $\{ f^{n_k} \left(y \right) \}$ and a natural number p such that $\left(f^{n_k} \left(y \right), x^* \right) \in E(G)$ for all $k \geq p$. We say that a graph G is an (H_f)-*graph* if $f^n \left(y \right) \in [x^*]_{\tilde{G}}$ for $n \geq 1$; then $r(f^n \left(y \right), x^*) \to 0$ (as $n \to \infty$), where $r(f^n \left(y \right), x^*) = \sum_{i=1}^{M_n} s^i d(z_{i-1}, z_i)$; $\{ z_i \}_{i=0}^{M_n}$, is a path from $f^n \left(y \right)$ to x^* in $\tilde{G}$.

Obviously every (C)-graph is a (C_f)-graph for any self-mapping f on X, but an example is given in [193] showing that the converse may not hold. Examples are also given in [193] showing that for a given f notions of (C_f)-graph and (H_f)-graph are independent even if f is identity map.

We now come to the main result of this section. Recall that a mapping $f : X \to X$ is called a *Picard operator* in the terminology of [171] if f has a unique fixed point $x^* \in X$ and $\lim_{n \to \infty} f^n \left(x \right) = x^*$ for each $x \in X$.

THEOREM 12.3. *Let (X, d) be a complete b-metric space and f be a b-(φ, G) contraction, where φ is b-comparison function. Assume d is continuous and there is z_0 in X for which $(z_0, f \left(z_0 \right))$ is an edge in $\tilde{G}$. Then the following assertions hold:*

   1. *If G is a (C_f)-graph, then f has a unique fixed point $x^* \in [z_0]_{\tilde{G}}$ and for any $y \in [z_0]_{\tilde{G}}$, $f^n \left(y \right) \to x^*$. Further if G is a weakly connected, then f is Picard operator.*
   2. *If G is a weakly connected (H_f)-graph, then f has a unique fixed point $x^* \in X$ and for any $y \in X$, $f^n \left(y \right) \to x^*$.*

PROOF. (1) It follows from Proposition 12.2 that $\{f^n(z_0)\}$ is a Cauchy sequence in X. Since X is complete, there exists $x^* \in X$ such that $\lim_{n\to\infty} f^n(z_0) = x^*$. Since $(f^n(z_0), f^{n+1}(z_0)) \in E(G)$ for all $n \in \mathbb{N}$, and G is a (C_f) graph, there exists a subsequence $\{f^{n_j}(z_0)\}$ of $\{f^n(z_0)\}$ and $p \in \mathbb{N}$ such that $(f^{n_j}(z_0), x^*) \in E(G)$ for all $j \geq p$. Observe that $(z_0, f(z_0), f^2(z_0), \cdots, f^{n_i}(z_0), \cdots, f^{n_p}(z_0), x^*)$ is a path in $\tilde{G}$. Therefore $x^* \in [z_0]_{\tilde{G}}$. Condition (12.3) now implies

$$d\left(f^{n_j+1}(z_0), f(x^*)\right) \leq \varphi\left(d\left(f^{n_j}(z_0), x^*\right)\right) < d\left(f^{n_j}(z_0), x^*\right) \text{ for each } j \geq n_0.$$

Since d is continuous, letting $j \to \infty$ we obtain $\lim_{j\to\infty} f^{n_j}(z_0) = f(x^*)$. Since $\{f^{n_j}(z_0)\}$ is a subsequence of $\{f^n(z_0)\}$, we conclude that $f(x^*) = x^*$. Finally, if $y \in [z_0]_{\tilde{G}}$, it follows from Lemma 12.1 that $\lim_{n\to\infty} f^n(y) = x^*$.

(2) Let G be a weakly connected $(H)_f$-graph. From Proposition 12.2, $f^n(z_0) \to x^*$ and thus $r(f^n(z_0), x^*) \to 0$ as $n \to \infty$. Now, for each $n \in \mathbb{N}$, let $\{y_i^n\}$ be a path in $\tilde{G}$ from $f^n(z_0)$ to x^* $(i = 0, 1, \cdots, M_n)$, with $y_0 = x^*$ and $y_{M_n}^n = f^n(z_0)$. Then

$$
\begin{aligned}
d(x^*, f(x^*)) &\leq s\left[d\left(x^*, f^{n+1}(z_0)\right) + d\left(f^{n+1}(z_0), f(x^*)\right)\right] \\
&\leq s\left[d\left(x^*, f^{n+1}(z_0)\right) + \sum_{i=1}^{M_n} s^i d\left(f\left(y_{i-1}^n\right), f\left(y_i^n\right)\right)\right] \\
&\leq s\left[d\left(x^*, f^{n+1}(z_0)\right) + \sum_{i=1}^{M_n} s^i \varphi\left(d\left(y_{i-1}^n, y_i^n\right)\right)\right] \\
&< s\left[d\left(x^*, f^{n+1}(z_0)\right) + \sum_{i=1}^{M_n} s^i d\left(y_{i-1}^n, y_i^n\right)\right] \\
&= s\left[d\left(x^*, f^{n+1}(z_0)\right) + r\left(f^n(z_0), x^*\right)\right].
\end{aligned}
$$

Letting $n \to \infty$, the above inequality yields $f(x^*) = x^*$. Let $y \in [z_0]_{\tilde{G}} = X$ be arbitrary. Then from Lemma 12.1 and Remark 12.2 it is easily seen that $\lim_{n\to\infty} f^n(y) = x^*$. $\square$

12.4. Strong b-Metric Spaces

It is easy to see that the distance function in a b-metric space need not be continuous. In fact if $\{q_n\} \subset X$ and if $\lim_{n\to\infty} q_n = q$, then for any $p \in X$ all that can be said is that

$$s^{-1}d(p, q) \leq \liminf_{n\to\infty} d(p, q_n) \leq \limsup_{n\to\infty} d(p, q_n) \leq sd(p, q).$$

In general $\lim_{n\to\infty} d(p, q_n) = d(p, q) \Leftrightarrow s = 1$. Indeed, open balls in such spaces need not be open sets. This prompts us to suggest a strengthening of the notion of b-metric spaces which remedies this defect.

DEFINITION 12.7. A semimetric space (X, d) is said to be a *strong b-metric space* (an *sb*-metric space for short) if there exists $s \geq 1$ such that for each $p, q, r \in X$

$$(12.7) \qquad\qquad d(p, q) \leq d(q, r) + s\, d(p, r).$$

PROPOSITION 12.3. *A semimetric space* (X, d) *is an sb-metric space if and only if there exists* $s \geq 1$ *such that for each* $p, q, r, t \in X$,

$$(12.8) \qquad\qquad |d(p, q) - d(r, t)| \leq s\,[d(p, r) + d(q, t)].$$

PROOF. Suppose (X, d) is an *sb*-metric space with constant $s \geq 1$. Then there exists $s \geq 1$ such that for all $p, q, r, t \in X$

$$\begin{aligned} d(p, q) &\leq d(p, r) + s\, d(q, r) \\ &\leq d(r, t) + s\, d(p, t) + s\, d(q, r) \end{aligned}$$

from which

$$d(p, q) - d(t, r) \leq s\,[d(p, t) + d(q, r)].$$

A similar argument shows that

$$d(t, r) - d(p, q) \leq s\,[d(t, p) + d(r, q)];$$

hence

$$|d(p, q) - d(t, r)| \leq s\,[d(p, r) + d(q, t)].$$

Thus an *sb*-metric space satisfies (12.8). The converse is trivial. $\qquad\square$

REMARK 12.3. *It is interesting to note that if* $s = 1$ *in inequality* (*12.8*), *then the resulting inequality is precisely the triangle inequality. This is because the relation* $|d(p, q) - d(r, t)| \leq d(p, r) + d(q, t)$ *implies* (*upon taking* $t = r$) $d(p, q) \leq d(p, r) + d(q, r)$. *Thus the triangle inequality holds. Conversely, if the triangle inequality is valid in* (X, d), *then*

$$\begin{aligned} |d(p, q) - d(r, t)| &= |d(p, q) - d(q, r) + d(q, r) - d(r, t)| \\ &\leq |d(p, q) - d(q, r)| + |d(q, r) - d(r, t)| \\ &\leq d(p, r) + d(q, t). \end{aligned}$$

REMARK 12.4. *While the concept of an sb-metric space is appealing on the surface, it would be nice to know whether there are interesting* (*natural*) *examples of such spaces.*

REMARK 12.5. *In view of Proposition 12.3, sb-metric spaces are precisely quasimetric spaces which satisfy condition* (*2.6*) *of Xia* [**218**].

To see that the distance function in an *sb*-metric space is continuous in the sense of Definition 1.3 we apply Proposition 12.3. Let $\{p_n\}, \{q_n\} \subseteq X$, and suppose

$$\lim_{n \to \infty} d(p_n, p) = 0 \text{ and } \lim_{n \to \infty} d(q_n, q) = 0.$$

Then (12.8) implies

$$|d(p, q) - d(p_n, q_n)| \leq s\,[d(p, p_n) + d(q, q_n)]$$

from which $\lim_{n\to\infty} d(p_n, q_n) = d(p, q)$. *An important consequence of this fact is that open balls are always open sets in an sb-metric space.*

While the triangle inequality is not always needed in metric fixed point theory, continuity of the distance function is extremely useful. Of course it would be possible to just assume one has a b-metric space with a continuous distance function. This is commonly done (see, e.g., [**31, 193**]). However codifying both facts with inequality (12.8) seems both more elegant and more in keeping with our "distance" approach. This formulation has other advantages as well. Notably, sb-metric spaces satisfy the relaxed polygonal inequality. This in turn assures that the Cauchy summation criterion for convergence of sequences holds. As we shall see, many fixed point theorems require only this latter fact.

PROPOSITION 12.4. *If a semimetric space (X, d) is an sb-metric space, then it is an s-relaxed$_p$ metric space.*

PROOF. Suppose X is an sb-metric space and let $\{p_n\} \subseteq X$. We assert that for any $n, j \in \mathbb{N}, j \geq 1$

$$(12.9) \qquad d(p_n, p_{n+j}) \leq d(p_n, p_{n+1}) + s \sum_{i=n+1}^{n+j-1} d(p_i, p_{i+1}).$$

The proof is by induction on j. Clearly (12.9) holds for $j = 1$ and, by definition of an sb-metric space, for $j = 2$. Assume that for $j \geq 2$ and $n \in \mathbb{N}$,

$$(12.10) \qquad d(p_n, p_{n+j}) \leq d(p_n, p_{n+1}) + s \sum_{i=n+1}^{n+j-1} d(p_i, p_{i+1}).$$

Then by (12.8)

$$|d(p_n, p_{n+j+1}) - d(p_n, p_{n+j})| \leq sd(p_{n+j}, p_{n+j+1}).$$

This along with the inductive assumption gives

$$\begin{aligned} d(p_n, p_{n+j+1}) &\leq d(p_n, p_{n+j}) + sd(p_{n+j}, p_{n+j+1}) \\ &\leq d(p_n, p_{n+1}) + s \sum_{i=n+1}^{n+j-1} d(p_i, p_{i+1}) + sd(p_{n+j}, p_{n+j+1}) \\ &= d(p_n, p_{n+1}) + s \sum_{i=n+1}^{n+j} d(p_i, p_{i+1}). \end{aligned}$$

This completes the induction.

At the same time, since $d(p_n, p_{n+1}) \leq sd(p_n, p_{n+1})$, this implies that

$$d(p_n, p_{n+j}) \leq s \sum_{i=n}^{n+j-1} d(p_i, p_{i+1}).$$

Taking $p_n = x$, $p_{n+j} = y$ and $x_k = p_{n+k}$, $k = 1, \cdots, j-1$, we have

$$d(x, y) \leq s[d(x, x_1) + d(x_1, x_2) + \cdots + d(x_{j-1}, y)].$$

□

PROPOSITION 12.5. *Let $\{p_n\}$ be a sequence in an sb-metric space and suppose*

$$\sum_{i=1}^{\infty} d(p_i, p_{i+1}) < \infty.$$

Then $\{p_n\}$ is a Cauchy sequence.

PROOF. This is immediate from the relaxed polygonal inequality. If $\varepsilon > 0$ there exists $N \in \mathbb{N}$ such that $n \geq N$ implies

$$\sum_{i=n}^{n+j} d(p_i, p_{i+1}) \leq \varepsilon.$$

Thus for all $j, n \in \mathbb{N}$ with n sufficiently large,

$$d(p_n, p_{n+j+1}) \leq s \sum_{i=n}^{n+j} d(p_i, p_{i+1}) \leq s\varepsilon.$$

□

12.5. Banach's Theorem in a Relaxed$_p$ Metric Space

The Cauchy convergence criterion yields a quick proof of the Banach contraction mapping theorem in s-relaxed$_p$ metric spaces. However, as we note above, Czerwik has shown that this theorem actually holds in a b-metric space.

THEOREM 12.4. *Let (X, d) be a complete s-relaxed$_p$ metric space, suppose $k \in (0, 1)$, and suppose $f : X \to X$ satisfies*

(12.11) $$d(f(x), f(y)) \leq kd(x, y)$$

for all $x, y \in X$. Then f has a unique fixed point x^*, and moreover the Picard iterates $\{f^n(x)\}$ converge to x^* for all $x \in X$.

PROOF. Define $\varphi : X \to \mathbb{R}$ by setting $\varphi(x) = (1-k)^{-1} d(x, f(x))$ for $x \in X$. Then

$$d(x, f(x)) - kd(x, f(x)) \leq d(x, f(x)) - d(f(x), f^2(x)).$$

Hence

$$\begin{aligned} d(x, f(x)) &\leq (1-k)^{-1}[d(x, f(x)) - d(f(x), f^2(x))] \\ &= \varphi(x) - \varphi(f(x)). \end{aligned}$$

Therefore

$$\sum_{i=0}^{\infty} d(f^i(x), f^{i+1}(x)) = \sum_{i=0}^{\infty} [\varphi(f^i(x)) - \varphi(f^{i+1}(x))] < \infty.$$

By the relaxed polygonal inequality $\{f^n(x)\}$ is a Cauchy sequence. Since X is complete $\{f^n(x)\}$ converges to a point $x^* \in X$, and since f is continuous, $f(x^*) = x^*$. Uniqueness follows from (12.11) and the fact that $d(x,y) = 0 \Leftrightarrow x = y$. $\square$

Theorem 12.4 extends readily to set-valued contraction. Let (X, d) be an sb-metric space, and let $\mathcal{CB}(X)$ be the collection of all nonempty bounded closed subsets of X. Define the Hausdorff distance H_{ab} on $\mathcal{CB}(X)$ in the usual way (see Sect. 9.7). The following is a generalization of Nadler's set-valued contraction mapping theorem in metric spaces. With the aid of Proposition 12.5 Nadler's original proof of [159] carries over with only minor change. We include the details.

12.6. Nadler's Theorem

THEOREM 12.5. *Let (X, d) be a complete s-relaxed$_p$ metric space, and let $\mathcal{CB}(X)$ be the collection of all nonempty bounded closed subsets of X endowed with the Hausdorff sb-metric H_{sb}. Let $k \in (0, 1)$ and suppose $T : X \to \mathcal{CB}(X)$ satisfies*

$$(12.12) \qquad H_{sb}(T(x), T(y)) \le kd(x, y)$$

for all $x, y \in X$. Then there exists $x^* \in X$ such that $x^* \in T(x^*)$.

PROOF. Select $x_0 \in X$ and $x_1 \in T(x_0)$. Since x_1 lies in an $H_{sb}(T(x_0), T(x_1))$ neighborhood of $T(x_1)$, there exists $x_2 \in T(x_1)$ such that

$$d(x_1, x_2) \le H_{sb}(T(x_0), T(x_1)) + k.$$

Similarly there exists $x_2 \in T(x_2)$ such that

$$d(x_2, x_3) \le H_{sb}(T(x_1), T(x_2)) + k^2.$$

Continuing in this manner one obtains a sequence $\{x_n\}$ with $x_{i+1} \in T(x_i)$ such that

$$\begin{aligned}
d(x_i, x_{i+1}) &\le H_{sb}(T(x_{i-1}), T(x_i)) + k^i \\
&\le kd(x_{i-1}, x_i) + k^i \\
&\le k\left[H_{sb}(T(x_{i-2}), T(x_{i-1})) + k^{i-1}\right] + k^i \\
&\le k^2 d(x_{i-2}, x_{i-1}) + 2k^i \\
&\le \cdots \\
&\le k^i d(x_0, x_1) + ik^i.
\end{aligned}$$

It follows that

$$\sum_{i=0}^{\infty} d(x_i, x_{i+1}) \le d(x_0, x_1) \sum_{i=0}^{\infty} k^i + \sum_{i=0}^{\infty} ik^i < \infty$$

so by the relaxed polygonal inequality $\{x_n\}$ is a Cauchy sequence. Since X is complete, there exists $x^* \in X$ such that $\lim_{n \to \infty} x_n = x^*$. By (12.12)

$$\lim_{n \to \infty} H_{sb}\left(T\left(x_n\right), T\left(x^*\right)\right) \leq k \lim_{n \to \infty} d\left(x_n, x^*\right) = 0.$$

Since $x_n \in T\left(x_{n-1}\right)$, $\lim_{n \to \infty} dist\left(x_n, T\left(x^*\right)\right) = 0$ and since $T\left(x^*\right)$ is closed, it follows that $x^* \in T\left(x^*\right)$. □

Remark. In [57] Czerwik obtains the same result as Theorem 12.5 for b-metric spaces, but with the further restriction that $k \leq s^{-1}$.

The proof of Ostrowski's stability result [165] also carries over to the sb-setting with only very minor modification of his original proof.

THEOREM 12.6. *Let* (X, d) *be a complete sb-metric space with* $s > 1$. *Let* $f : X \to X$ *be a contraction mapping with Lipschitz constant* $k \in (0, 1)$, *and suppose* x^* *is the unique fixed point of* f. *Let* $\{\varepsilon_n\}$ *be a sequence of positive numbers for which* $\lim_{n \to \infty} \varepsilon_n = 0$. *Let* $y_0 \in X$ *and suppose* $\{y_n\} \subset X$ *satisfies*

$$d\left(y_{n+1}, f\left(y_n\right)\right) \leq \varepsilon_n, \quad n \in \mathbb{N}.$$

Then $\lim_{n \to \infty} y_n = x^*$.

PROOF. Let $m \in \mathbb{N}$. Then

$$
\begin{aligned}
d\left(f^{m+1}\left(y_0\right), y_{m+1}\right) &\leq d\left(f\left(f^m\left(y_0\right)\right), f\left(y_m\right)\right) + sd\left(f\left(y_m\right), y_{m+1}\right) \\
&\leq kd\left(f^m\left(y_0\right), y_m\right) + s\varepsilon_m \\
&\leq kd\left(f\left(f^{m-1}\left(y_0\right)\right), f\left(y_{m-1}\right)\right) \\
&\quad + skd\left(f\left(y_{m-1}\right), y_m\right) + s\varepsilon_m \\
&\leq k^2 d\left(f^{m-1}\left(y_0\right), y_{m-1}\right) + sk\varepsilon_{m-1} + s\varepsilon_m \\
&\vdots \\
&\leq \sum_{i=0}^{m} k^{m-i} s\varepsilon_i.
\end{aligned}
$$

Therefore

$$
\begin{aligned}
d\left(y_{m+1}, x^*\right) &\leq d\left(y_{m+1}, f^{m+1}\left(y_0\right)\right) + sd\left(f^{m+1}\left(y_0\right), x^*\right) \\
&\leq \sum_{i=0}^{m} k^{m-i} s\varepsilon_i + sd\left(f^{m+1}\left(y_0\right), x^*\right).
\end{aligned}
$$

Now let $\varepsilon > 0$. Since $\lim_{n \to \infty} \varepsilon_n = 0$, there exists $N \in \mathbb{N}$ such that for $m > N$, $s\varepsilon_m \leq \varepsilon$. Thus

$$
\begin{aligned}
\sum_{i=0}^{m} k^{m-i} s\varepsilon_i &= s\sum_{i=0}^{N} k^{m-i} \varepsilon_i + \sum_{i=N+1}^{m} k^{m-i} s\varepsilon_i \\
&\leq k^{m-N} \sum_{i=0}^{N} k^{N-i} s\varepsilon_i + \varepsilon \sum_{i=N+1}^{m} k^{m-i}.
\end{aligned}
$$

Hence $\lim_{m\to\infty} \sum_{i=0}^{m} k^{m-i} s\varepsilon_i \leq \varepsilon \sum_{i=0}^{\infty} k^i$, and since $\varepsilon > 0$ is arbitrary, it follows that $\lim_{m\to\infty} \sum_{i=0}^{m} k^{m-i} s\varepsilon_i = 0$. Since $\lim_{m\to\infty} sd\left(f^{m+1}(y_0), x^*\right) = 0$, we conclude that $\lim_{m\to\infty} y_{m+1} = x^*$. $\qquad\square$

It is only fair to point out that some results seem to require the full use of the triangle inequality. In this connection we mention an interesting extension of Nadler's theorem due to Dontchev and Hager [62] using the concept of the excess from a set A to a set B in a metric space. Let (X, d) be a metric space, and for $A \subseteq X$ and $x \in X$, set

$$dist(x, A) = \inf\{d(x, a) : a \in A\}.$$

The *excess* δ from A to the set $B \subseteq X$ is given by

$$\delta(B, A) = \sup\{dist(x, A) : x \in B\}.$$

The following generalization of Nadler's theorem is used in [62] to prove an inverse mapping theorem for set-valued mappings T from a complete metric space X to a linear space Y with a translation invariant metric. We prove the original version of this theorem here. Another proof is given in the recent paper [20], where it is pointed out that this "local" version does in fact include Nadler's original theorem.

THEOREM 12.7. *Let (X, d) be a complete metric space and suppose T maps X into the nonempty closed subsets of X. Let $x_0 \in X$ and suppose $r \in \mathbb{R}^+$ and $k \in [0, 1)$ satisfy*
  (a) $dist(x_0, T(x_0)) < r(1 - k)$;
  (b) $\delta(T(x_1) \cap B(x_0; r), T(x_2)) \leq kd(x_1, x_2)$ *for all $x_1, x_2 \in B(x_0; r)$.*
Then T has a fixed point in $B(x_0; r)$.

PROOF ([62]). By assumption (a) there exists $x_1 \in T(x_0)$ such that $d(x_1, x_0) < r(1 - k)$. Proceeding by induction, suppose that there exists $x_{j+1} \in T(x_j) \cap B(x_0; r)$ such that

$$d(x_{j+1}, x_j) < r(1 - k)k^j$$

for $j = 1, 2, \cdots, n - 1$. By assumption (b)

$$dist(x_n, T(x_n)) \leq \delta(T(x_{n-1}) \cap B(x_0; r), T(x_n)) \leq kd(x_n, x_{n-1}) < r(1 - k)k^n.$$

This implies that there exists $x_{n+1} \in T(x_n)$ such that

$$d(x_{n+1}, x_n) < r(1 - k)k^n.$$

However (and here we make full use of the triangle inequality)

$$d(x_{n+1}, x_0) \leq \sum_{j=0}^{n} d(x_{j+1}, x_j) < r(1 - k)\sum_{j=0}^{n} k^j < r.$$

Hence $x_{j+1} \in T(x_j) \cap B(x_0; r)$. This completes the induction.
   For $n > m$ we now have

$$d(x_n, x_m) \leq \sum_{j=m}^{n-1} d(x_{j+1}, x_j) < r(1 - k)\sum_{j=m}^{n-1} k^j < rk^m.$$

Thus $\{x_j\}$ is a Cauchy sequence which converges to some $x^* \in B(x_0; r)$. By assumption (b)

$$dist\,(x_n, T\,(x^*)) \leq \delta\,(T\,(x_{n-1}) \cap B\,(x_0; r), T\,(x^*)) \leq kd\,(x^*, x_{n-1}).$$

The triangle inequality now implies that

$$dist\,(x^*, T\,(x^*)) \leq d\,(x^*, x_n) + dist\,(x_n, T\,(x^*)) \leq d\,(x_n, x^*) + kd\,(x_{n-1}, x^*).$$

Since the latter term approaches 0 as $n \to \infty$, and since $T\,(x^*)$ is closed, it follows that $x^* \in T\,(x^*)$.   □

We conclude with two questions.

QUESTION. Does Theorem 12.7 hold under the weaker sb-metric assumption?

QUESTION. Is every sb-metric space X dense in a complete sb-metric space X'? If so, then every contraction mapping $f : X \to X$ extends to a contraction mapping $f' : X' \to X'$ which has a unique fixed point. Ostrowski's theorem then would provide a method for approximating this fixed point.

12.7. Caristi's Theorem in sb-Metric Spaces

The following is Theorem 2.4 of [**31**]. It is derived from a version of Ekeland's variational principle in b-metric spaces.

THEOREM 12.8. *Let (X, d) be a complete b-metric space, (with $s > 1$) such that the b-metric d is continuous and let $\psi : X \to \mathbb{R}$ be lower semicontinuous. Suppose $f : X \to X$ satisfies*

$$(12.13) \qquad\qquad d\,(u, v) + sd\,(u, f\,(u)) \geq d\,(f\,(u), v)$$

and

$$(12.14) \qquad\qquad \frac{s^2}{s-1} d\,(u, f\,(u)) \leq \psi\,(u) - \psi\,(f\,(u))$$

for all $u, v \in X$. Then f has a fixed point.

This quickly yields Caristi's theorem for sb-metric spaces.

COROLLARY 12.1. *Let (X, d) be a complete sb-metric space (with $s > 1$) and let $\varphi : X \to \mathbb{R}$ be lower semicontinuous and bounded below. Suppose $f : X \to X$ satisfies*

$$d\,(x, f\,(x)) \leq \varphi\,(x) - \varphi\,(f\,(x))$$

for all $x \in X$. Then f has a fixed point.

PROOF. Continuity of the distance functions comes from the fact that d is an sb-metric. Also, taking $q = t$ in (12.8) we obtain

$$|d(p,t) - d(r,t)| \leq sd(p,r)$$

for each $p, r, t \in X$. Thus

$$d(r,t) + sd(p,r) \geq d(p,t)$$

for each $p, r, t \in X$, and it follows upon taking $p = f(u)$, $t = v$, and $r = u$, that

$$d(u,v) + sd(u, f(u)) \geq d(f(u), v)$$

for each $u, v \in X$, so (12.13) holds. Finally, taking $\psi = \dfrac{s^2}{s-1}\varphi$, we obtain (12.14). $\square$

REMARK 12.6. *We do not know whether Caristi's theorem fully extends to b-metric spaces. However, as we show in Chap. 14, it does extend to partial metric spaces.*

12.8. The Metric Boundedness Property

We now discuss the relation between b-metric spaces, relaxed$_p$ metric spaces, and metric boundedness. Recall that a semimetric space (X, d) is s-metric bounded (for $s \geq 1$) if there is a metric ρ on X such that for all $x, y \in X$,

$$\rho(x,y) \leq d(x,y) \leq s\rho(x,y).$$

THEOREM 12.9 ([**78**]). *Let (X, d) be a semimetric space. Then (X, d) is an s-relaxed$_p$ metric if and only if (X, d) is s-metric bounded.*

PROOF. ($\Rightarrow$) Assume (X, d) is an s-relaxed$_p$ metric. Define ρ by taking

$$(12.15) \qquad \rho(x,y) = \min_{\ell} \min_{\{x_0, x_1, \cdots, x_\ell : x_0 = x ; x_\ell = y\}} \sum_{i=0}^{\ell-1} d(x_i, x_{i+1})$$

for each $x, y \in X$.

We first show that ρ is a metric. Since $d(x,x) = 0$ it follows from (12.15) that $\rho(x,x) = 0$. Now suppose $x, y \in X$ with $x \neq y$. By the relaxed polygonal inequality for s we know that in expression (12.15)

$$d(x,y) \leq s \sum_{i=0}^{\ell-1} d(x_i, x_{i+1}).$$

Therefore

$$\rho(x,y) \geq \frac{1}{s}d(x,y),$$

and since $d(x, y) > 0$ it follows that $\rho(x, y) > 0$. Symmetry of ρ is immediate from the definition. Finally ρ satisfies the triangle inequality because for any $x, y, z \in X$,

$$
\begin{aligned}
\rho(x, y) &= \min_{\ell} \min_{\{x_0, x_1, \cdots, x_\ell : x_0 = x; x_\ell = y\}} \sum_{i=0}^{\ell-1} d(x_i, x_{i+1}) \\
&\leq \min_{\ell_1} \min_{\{y_0, y_1, \cdots, y_{\ell_1} : y_0 = x; y_{\ell_1} = z\}} \sum_{i=0}^{\ell_1 - 1} d(y_i, y_{i+1}) \\
&\quad + \min_{\ell_2} \min_{\{z_0, z_1, \cdots, z_{\ell_2} : z_0 = z; z_{\ell_2} = y\}} \sum_{i=0}^{\ell_2 - 1} d(z_i, z_{i+1}) \\
&= \rho(x, z) + \rho(z, y).
\end{aligned}
$$

To see that (X, d) is s-metric bounded, by (12.15) it follows easily that $\rho(x, y) \leq d(x, y)$, and also by (12.15) and the relaxed polygonal inequality, $d(x, y) \leq s\rho(x, y)$.

($\Leftarrow$) Now assume (X, d) is s-metric bounded. Then there is a metric ρ on X such that for all $x, y \in X$,

$$\rho(x, y) \leq d(x, y) \leq s\rho(x, y).$$

Therefore $d(x, x) = 0$. If $x \neq y$, then $d(x, y) \geq \rho(x, y) > 0$. To see that d satisfies the relaxed polygonal inequality, let $x, x_1, \cdots, x_{n-1}, y \in X$. Then

$$
\begin{aligned}
d(x, y) &\leq s\rho(x, y) \\
&\leq s[\rho(x, x_1) + \rho(x_1, x_2) + \cdots + \rho(x_{n-1}, y)] \text{ (since } \rho \text{ is a metric)} \\
&\leq s[d(x, x_1) + d(x_1, x_2) + \cdots + d(x_{n-1}, y)] \text{ (since } \rho(\cdot, \cdot) \leq d(\cdot, \cdot)).
\end{aligned}
$$

Since d is semimetric by assumption, it follows that (X, d) is an s-relaxed$_p$ metric. $\qquad \square$

Having shown that the notions of an s-relaxed$_p$ metric space and metric boundedness are equivalent, we compare these concepts to the concept of a b-metric space. Every s-relaxed$_p$ metric space is a b-metric space by definition. We see below that the converse is not true.

THEOREM 12.10 ([**78**]). *There is a b-metric space that is not an s-relaxed$_p$ metric for any s.*

PROOF. Let $X = [0, 1]$ and define d on X by setting $d(x, y) = (x - y)^2$. Clearly (X, d) is a semimetric space. To see that (X, d) is a b-metric space, let $x, y, z \in X$ and set $\alpha = d(x, z)$, $\beta = d(z, y)$ and $\gamma = d(x, y)$. Then

$\sqrt{\gamma} \leq \sqrt{\alpha} + \sqrt{\beta}$ since $|\cdot - \cdot|$ is the standard metric on X. Therefore $\gamma \leq \alpha + \beta + 2\sqrt{\alpha\beta}$. But $\sqrt{\alpha\beta} \leq \dfrac{\alpha + \beta}{2}$. It follows that

$$d(x, y) \leq 2\left[d(x, z) + d(z, y)\right].$$

So d is b-metric with constant 2.

Now let n be an arbitrary positive integer and let $x_i = \dfrac{i}{n}$ for $1 \leq i \leq n-1$. Then

$$d(0, x_1) + d(x_1, x_2) + \cdots + d(x_{n-1}, 1) = n\left(\frac{1}{n}\right)^2 = \frac{1}{n}.$$

Since n is arbitrary it is clear that there can be no constant s such that

$$d(0, 1) \leq s\left[d(0, x_1) + d(x_1, x_2) + \cdots + d(x_{n-1}, 1)\right].$$

$\square$

CHAPTER 13

Generalized Metric Spaces

13.1. Introduction

We now turn to a concept introduced by A. Branciari. This class of spaces has received significant attention lately, although at this point it remains unclear whether the concept has any significant applications.

DEFINITION 13.1 ([35]). Let X be a nonempty set and $d : X \times X \to [0, \infty)$ a mapping such that for all $x, y \in X$ and all distinct points $u, v \in X$, each distinct from x and y :

   (i) $d(x, y) = 0 \Leftrightarrow x = y$;
   (ii) $d(x, y) = d(y, x)$;
   (iii) $d(x, y) \leq d(x, u) + d(u, v) + d(v, y)$ (quadrilateral inequality).

Then X is called a *generalized metric space* (g.m.s.).

PROPOSITION 13.1. *If (X, d) is a generalized metric space which satisfies Wilson's Axiom III (see Chap. 1), then the distance function is continuous at distinct points.*

PROOF. Suppose $\{p_n\}, \{q_n\} \subseteq X$ satisfy

$$\lim_{n \to \infty} d(p_n, p) = 0 \text{ and } \lim_{n \to \infty} d(q_n, q) = 0,$$

where $p \neq q$. Also assume that for n arbitrarily large, $p_n \neq p$ and $q_n \neq q$. In view of Axiom III, we may also assume that for n sufficiently large, $p_n \neq q_n$. Then

$$d(p, q) \leq d(p, p_n) + d(p_n, q_n) + d(q_n, q)$$

and

$$d(p_n, q_n) \leq d(p_n, p) + d(p, q) + d(q, q_n).$$

Together these inequalities imply

$$\liminf_{n \to \infty} d(p_n, q_n) \geq d(p, q) \geq \limsup_{n \to \infty} d(p_n, q_n).$$

Thus $\lim_{n \to \infty} d(p_n, q_n) = d(p, q)$. $\square$

Proposition 2 of [129] asserts that the distance function is continuous. However, to get full continuity of d it would be necessary to show that if $\lim_{n \to \infty} d(p_n, p) = 0$ and $\lim_{n \to \infty} d(q_n, p) = 0$, then $\lim_{n \to \infty} d(p_n, q_n) = 0$. While this is a trivial consequence of the triangle inequality, the quadrilateral

© Springer International Publishing Switzerland 2014
W. Kirk, N. Shahzad, *Fixed Point Theory in Distance Spaces*,
DOI 10.1007/978-3-319-10927-5_13

inequality alone is not strong enough to assure this. However the following observation shows that the quadrilateral inequality implies a weaker but useful form of distance continuity. (This is a special case of Proposition 1 of [211]. Also see Lemma 1.10 of [102].)

PROPOSITION 13.2. *Suppose $\{q_n\}$ is a Cauchy sequence in a generalized metric space X and suppose $\lim_{n\to\infty} d(q_n, q) = 0$. Then $\lim_{n\to\infty} d(p, q_n) = d(p, q)$ for all $p \in X$. In particular, $\{q_n\}$ does not converge to p if $p \neq q$.*

PROOF. If $q_n = p$ for arbitrarily large n, it must be the case that $p = q$. So we may also assume that eventually $p \neq q_n$. Also eventually $q_n \neq q$; otherwise the result is trivial. So by passing to a subsequence we may assume that $q_n \neq q_m \neq q$ and $q_n \neq q_m \neq p$ for all $n, m \in \mathbb{N}$ with $n \neq m$. Then by the quadrilateral inequality,

$$d(p, q) \leq d(p, q_n) + d(q_n, q_{n+1}) + d(q_{n+1}, q)$$

and

$$d(p, q_n) \leq d(p, q) + d(q, q_{n+1}) + d(q_{n+1}, q_n).$$

Since $\{q_n\}$ is a Cauchy sequence, $\lim_{n\to\infty} d(q_n, q_m) = 0$. Therefore

$$\limsup_{n\to\infty} d(p, q_n) \leq d(p, q) \leq \liminf_{n\to\infty} d(p, q_n).$$

□

We now come to Branciari's extension of Banach's contraction mapping theorem. Although in his proof Branciari makes the erroneous assertion that a generalized metric space is a Hausdorff topological space with a neighborhood basis given by

$$\mathbb{B} = \{B(x; r) : x \in S, r \in \mathbb{R}^+ \backslash \{0\}\},$$

with the aid of Proposition 13.2, Branciari's proof carries over with only minor change. The assertion in [194] that the space needs to be Hausdorff is superfluous, a fact first noted by Turinici in [211].

THEOREM 13.1 ([35]). *Let (X, d) be a complete generalized metric space, and suppose the mapping $f : X \to X$ satisfies $d(f(x), f(y)) \leq kd(x, y)$ for all $x, y \in X$ and fixed $k \in (0, 1)$. Then f has a unique fixed point x^*, and $\lim_{n\to\infty} f^n(x) = x^*$ for each $x \in X$.*

PROOF. It is possible to prove this theorem by following the proof given by Branciari up to the point of showing that $\{f^n(x)\}$ is a Cauchy sequence for each $x \in X$. We give the details. Let $x \in X$ and consider the sequence $\{f^n(x)\}$. If $f^i(x) = x$ for some $i \in \mathbb{N}$, then

$$d(x, f(x)) = d\left(f^i(x), f^{i+1}(x)\right) \leq k^i d(x, f(x))$$

and it follows that $f(x) = x$. Thus either f has a fixed point or $f^n(x) \neq f^m(x)$ if $m \neq n$.

We assert that for each $y \in X$

(a) $d\left(y, f^{2n}\left(y\right)\right) \leq \sum_{i=0}^{2n-3} k^i d\left(y, f\left(y\right)\right) + k^{2n-2} d\left(y, f^2\left(y\right)\right)$ for $n = 2, 3, 4, \cdots$;

(b) $d\left(y, f^{2n+1}\left(y\right)\right) \leq \sum_{i=0}^{2n} k^i d\left(y, f\left(y\right)\right)$ for $n = 0, 1, 2, \cdots$.

The proof is by mathematical induction. To see that (a) is true, for $n = 2$ one has

$$
\begin{aligned}
d\left(y, f^4\left(y\right)\right) &\leq d\left(y, f\left(y\right)\right) + d\left(f\left(y\right), f^2\left(y\right)\right) + d\left(f^2\left(y\right), f^4\left(y\right)\right) \\
&\leq d\left(y, f\left(y\right)\right) + kd\left(y, f\left(y\right)\right) + k^2 d\left(y, f^2\left(y\right)\right).
\end{aligned}
$$

Now let $n_0 \in \mathbb{N}$ and suppose that (a) holds for all $n \in \mathbb{N}$ such that $2 \leq n \leq n_0$. Then

$$
\begin{aligned}
d\left(y, f^{2n_0+2}\left(y\right)\right) &\leq d\left(y, f\left(y\right)\right) + d\left(f\left(y\right), f^2\left(y\right)\right) + d\left(f^2\left(y\right), f^{2n_0+2}\left(y\right)\right) \\
&\leq d\left(y, f\left(y\right)\right) + kd\left(y, f\left(y\right)\right) + k^2 d\left(y, f^{2n_0}\left(y\right)\right) \\
&\leq d\left(y, f\left(y\right)\right) + kd\left(y, f\left(y\right)\right) \\
&\quad + k^2 \left[\sum_{i=0}^{2n_0-3} k^i d\left(y, f\left(y\right)\right) + k^{2n_0-2} d\left(y, f^2\left(y\right)\right) \right] \\
&= \sum_{i=0}^{2n_0-1} k^i d\left(y, f\left(y\right)\right) + k^{2n_0} d\left(y, f^2\left(y\right)\right).
\end{aligned}
$$

To see that (b) is true, for $n = 0$ one has $d\left(y, f\left(y\right)\right) = d\left(y, f\left(y\right)\right)$. Now suppose (b) holds for some $n_0 \in \mathbb{N}$ and all $n \in \mathbb{N}$ with $0 \leq n \leq n_0$. Then for $n_0 + 1$,

$$
\begin{aligned}
d\left(y, f^{2n_0+3}\left(y\right)\right) &\leq d\left(y, f\left(y\right)\right) + d\left(f\left(y\right), f^2\left(y\right)\right) + d\left(f^2\left(y\right), f^{2n_0+3}\left(y\right)\right) \\
&\leq d\left(y, f\left(y\right)\right) + kd\left(y, f\left(y\right)\right) + k^2 d\left(y, f^{2n_0+1}\left(y\right)\right) \\
&\leq d\left(y, f\left(y\right)\right) + kd\left(y, f\left(y\right)\right) \\
&\quad + k^2 \sum_{i=0}^{2n_0} k^i d\left(y, f\left(y\right)\right) \\
&= \sum_{i=0}^{2n_0+2} k^i d\left(y, f\left(y\right)\right).
\end{aligned}
$$

It now follows that for all $n, m \in \mathbb{N}$

$$
\begin{aligned}
d\left(f^n\left(x\right), f^{n+2m}\left(x\right)\right) &\leq k^n d\left(x, f^{2m}\left(x\right)\right) \\
&\leq k^n \sum_{i=0}^{2m-2} k^i \max\left\{d\left(x, f\left(x\right)\right), d\left(x, f^2\left(x\right)\right)\right\} \\
&\leq \frac{k^n}{1-k} \max\left\{d\left(x, f\left(x\right)\right), d\left(x, f^2\left(x\right)\right)\right\}.
\end{aligned}
$$

Also

$$d\left(f^n\left(x\right), f^{n+2m+1}\left(x\right)\right) \le k^n d\left(x, f^{2m+1}\left(x\right)\right)$$

$$\le k^n \sum_{i=0}^{2m} k^i \max\left\{d\left(x, f\left(x\right)\right), d\left(x, f^2\left(x\right)\right)\right\}$$

$$\le \frac{k^n}{1-k} \max\left\{d\left(x, f\left(x\right)\right), d\left(x, f^2\left(x\right)\right)\right\}.$$

Thus for all $n, m \in \mathbb{N}$

$$d\left(f^n\left(x\right), f^m\left(x\right)\right) \le \frac{k^n}{1-k} \max\left\{d\left(x, f\left(x\right)\right), d\left(x, f^2\left(x\right)\right)\right\}.$$

Therefore $\{f^n\left(x\right)\}$ is a Cauchy sequence, and by completeness of X there exists $x^* \in X$ such that $\lim_{n\to\infty} f^n\left(x\right) = x^*$. But $\lim_{n\to\infty} d\left(f^{n+1}\left(x\right), f\left(x^*\right)\right) \le k \lim_{n\to\infty} d\left(f^n\left(x\right), x^*\right) = 0$, so

$$\lim_{n\to\infty} f^{n+1}x = f\left(x^*\right).$$

In view of Proposition 13.2, $f\left(x^*\right) = x^*$.   □

13.2. Caristi's Theorem in Generalized Metric Spaces

We begin with an examination of an easy proof of Caristi's original theorem in a complete metric space to illustrate why it fails in a generalized metric space. This proof is based on Zorn's Lemma (in contrast to the more constructive and more general approaches discussed earlier in Chap. 2).

THEOREM 13.2 (Caristi). *Let (X, d) be a complete metric space. Let $f : X \to X$ a mapping, and $\varphi : X \to \mathbb{R}^+$ a lower semicontinuous function. Suppose*

(13.1) $$d\left(x, f\left(x\right)\right) \le \varphi\left(x\right) - \varphi\left(f\left(x\right)\right), \qquad x \in X.$$

Then f has a fixed point.

PROOF. Introduce the Brøndsted partial order on X by setting $x \preceq y \Leftrightarrow d\left(x, y\right) \le \varphi\left(x\right) - \varphi\left(y\right)$. Let I be a totally ordered set and let $\{x_\gamma\}_{\gamma \in I}$ be a chain in $(X, \preceq)$. Then $\alpha \le \beta \Rightarrow x_\alpha \preceq x_\beta \Leftrightarrow d\left(x_\alpha, x_\beta\right) \le \varphi\left(x_\alpha\right) - \varphi\left(x_\beta\right)$. Therefore $\{\varphi\left(x_\gamma\right)\}_{\gamma \in I}$ is decreasing. Since φ is bounded below, $\lim_\gamma \varphi\left(x_\gamma\right) = r$. This implies $\lim_{\alpha,\beta} d\left(x_\alpha, x_\beta\right) = 0$; hence $\{x_\gamma\}_{\gamma \in I}$ is a Cauchy net. Since X is complete, there exists $x \in X$ such that $\lim_\gamma x_\gamma = x$. Thus for $\alpha \in I$,

$$d\left(x_\alpha, x\right) = \lim_\gamma d\left(x_\alpha, x_\gamma\right)$$

$$\le \lim_\gamma \left(\varphi\left(x_\alpha\right) - \varphi\left(x_\gamma\right)\right)$$

$$= \varphi\left(x_\alpha\right) - r$$

$$\le \varphi\left(x_\alpha\right) - \varphi\left(x\right).$$

Therefore $x_\alpha \preceq x$ for each $\alpha \in I$, so x is an upper bound for the chain $\{\varphi(x_\gamma)\}_{\gamma \in I}$. By Zorn's Lemma, $(X, \preceq)$ has a maximal element x^*. But condition (13.1) implies $x^* \preceq f(x^*)$, so it must be the case that $x^* = f(x^*)$. $\square$

The above argument fails in the setting of a generalized metric space because it is not possible to conclude that $(X, \preceq)$ is transitive in such a space. In a metric space, transitivity follows directly from the triangle inequality.

The assertion in [129] that Caristi's Theorem holds in generalized metric spaces is based on the false assertion that if $\{p_n\}$ is a sequence in a generalized metric space (X, d) which satisfies $\sum_{i=1}^\infty d(p_i, p_{i+1}) < \infty$, then $\{p_n\}$ is a Cauchy sequence. As we have seen in Proposition 12.5 this property is valid in sb-metric spaces, and with this additional assumption it is likely Caristi's theorem holds in a generalized metric space. However, as the following example shows, generalized metric spaces do not enjoy this Cauchy criterion. This example is a modification of Example 1 of [101]. See also [129] for details.

EXAMPLE 13.1. *Let* $X := \mathbb{N}$, *and define the function* $d : \mathbb{N} \times \mathbb{N} \to \mathbb{R}$ *by putting, for all* $m, n \in \mathbb{N}$,

$$d(n+1, n) = d(n, n+1) := \frac{1}{2^n};$$

$$d(n, n) = 0;$$

$$d(n, m) = d(m, n) := 1 \text{ if } m > n \text{ and } m - n \text{ is even;}$$

$$d(n, m) = d(m, n) := \sum_{i=n}^m d(i, i+1) \text{ if } m > n \text{ and } m - n \text{ is odd.}$$

To see that (X, d) *is a generalized metric space, let* $x_n = n$ $(n \in \mathbb{N})$, *suppose* $m, n \in \mathbb{N}$ *with* $m > n$ *and suppose* $p, q \in \mathbb{N}$ *are distinct with* $q > p$ *and* $p \neq n$ *and* $q \neq m$. *We now show that*

(13.2) $d(x_n, x_m) \leq d(x_n, x_p) + d(x_p, x_q) + d(x_q, x_m).$

If one of the three numbers $|n - p|$, $q - p$ *or* $|q - m|$ *is even, then, since*

$$d(x_n, x_m) \leq 1,$$

clearly (13.2) holds. If all the three numbers are odd, then, since $m - n = (m - q) + (q - p) + (p - n)$, *it follows that* $m - n$ *is odd and*

$$d(x_n, x_m) = \sum_{i=n}^m d(x_i, x_{i+1}).$$

There are four cases to consider:
  (i) $n < p < q < m$
  (ii) $p < n < q < m$
  (iii) $n < p < m < q$
  (iv) $p < n < m < q$

If (i) *holds, then*

$$d(x_n, x_m) = \sum_{i=n}^{m} d(x_i, x_{i+1})$$

$$= \sum_{i=n}^{p} d(x_i, x_{i+1}) + \sum_{i=p}^{q} d(x_i, x_{i+1}) + \sum_{i=q}^{m} d(x_i, x_{i+1})$$

$$= d(x_n, x_p) + d(x_p, x_q) + d(x_q, x_m).$$

In the other three cases

$$d(x_n, x_m) < d(x_n, x_p) + d(x_p, x_q) + d(x_q, x_m).$$

Therefore (X, d) *is a generalized metric space. Also* (X, d) *is complete because any Cauchy sequence in* X *must eventually be constant. Notice that if* $m, n \in \mathbb{N}$ *and* $m > n$, *then in order for* $d(x_n, x_m)$ *to be small for large* n, $m - n$ *must be odd, because if* $m - n$ *is even,* $d(x_n, x_m) = 1$. *However if* $\{x_{n_k}\}$ *is a subsequence of* $\{x_n\}$, *then* $|n_i - n_j|$ *cannot be odd for all sufficiently large* i, j. *(Suppose* $n_i > n_j > n_k$. *If* $n_i - n_j$ *is odd and if* $n_j - n_k$ *is odd, then* $n_i - n_k$ *is even.) Thus* $d(x_{n_i}, x_{n_j}) = 1$ *for arbitrarily large* i, j. *On the other hand, the Cauchy summation criterion fails, because* $\sum_{i=1}^{\infty} d(x_i, x_{i+1}) < \infty$, *and clearly* $\{x_n\}$ *is not a Cauchy sequence.*

REMARK 13.1. *Caristi's Theorem fails in the above example. Let* $f(n) = n + 1$ *for* $n \in \mathbb{N}$, *and define* $\varphi : \mathbb{N} \to \mathbb{R}$ *by setting* $\varphi(n) = \dfrac{2}{n}$. *Obviously* f *has no fixed points and, because the space is discrete,* φ *is continuous. On the other hand,* f *satisfies Caristi's condition:*

$$d(n, f(n)) \le \varphi(n) - \varphi(f(n)).$$

To see this, we need to show that

$$\frac{1}{2^n} \le \frac{2}{n} - \frac{2}{n+1} = \frac{2}{n(n+1)}.$$

This is equivalent to the assertion that

(13.3) $$2^{n+1} \ge n(n+1).$$

The proof is by induction. If $n = 1$,

$$2^2 = 4 > 2,$$

and for $n = 2$,

$$2^3 > 2 \cdot 3.$$

Assume (13.3) *holds for* $n \in \mathbb{N}$, $n \ge 2$. *Then*

$$\begin{aligned}
2^{n+2} &= 2(2^{n+1}) \\
&\ge 2n(n+1) \\
&= (n+n)(n+1) \\
&\ge (n+1)(n+2).
\end{aligned}$$

13.3. Multivalued Mappings in Generalized Metric Spaces

In the study of metric fixed point theory it is customary to investigate multivalued analogs of theorems first established for single-valued mappings. In view of Theorem 13.1 it is appropriate to see if Nadler's theorem for multivalued contraction mappings holds in generalized metric spaces. The starting point entails the Hausdorff metric H defined on the family of nonempty bounded closed subsets $\mathcal{CB}(X)$ of a given metric space (X, d) (see Sect. 9.7). It is well known that H is a metric on $\mathcal{CB}(X)$, and that $(\mathcal{CB}(X), H)$ is complete if and only if (X, d) is complete. However Kadelburg and Radenović have noted in [104] that an analogous construction is not possible in a generalized metric space.

EXAMPLE 13.2 ([104]). *Let $X = \{a, b, c\}$ and let $d : X \times X \to \mathbb{R}$ be defined as follows: $d(a, b) = 4$; $d(a, c) = d(b, c) = 1$; $d(x, x) = 0$ for $x \in X$, and $d(x, y) = d(y, x)$ for $x, y \in X$. The triangle inequality only need be checked when $x = y$ in which case it is trivial. Thus (X, d) is a generalized metric space which is obviously not a metric space.*
Now let H be the Hausdorff metric on $\mathcal{CB}(X)$, and consider the quadrilateral $(\{a\}, \{b\}, \{a, c\}, \{c\})$. It is easy to see that

$$H(\{a\}, \{b\}) = 4 > 1+1+1 = H(\{a\}, \{a, c\})+H(\{a, c\}, \{c\})+H(\{c\}, \{b\}).$$

Hence the quadrilateral inequality is not satisfied, so $(\mathcal{CB}(X), H)$ is not a generalized metric space.

CHAPTER 14

Partial Metric Spaces

14.1. Introduction

The topic of this section has its origins in the study of theoretical computer science. In 1992, S.G. Matthews [149] introduced the notion of a "partial metric space" with the aim of providing a quantitative mathematical model suitable for programming verification. See [150, 200] for further discussion. Among other things, Matthews proved a partial metric version of the celebrated Banach fixed point theorem which has become an appropriate quantitative fixed point technique to capture the meaning of recursive denotational specifications in programming languages. This is a class of distance spaces for which the triangle inequality is strengthened but for which Wilson's Axiom I (see Chap. 1) is relaxed. Thus these spaces are neither metric spaces nor semimetric spaces.

We begin with the relevant definitions. A *partial metric* [149] on a set X is a function $\rho : X \times X \to \mathbb{R}^+$ such that for all $x, y, z \in X$:

   (i) $x = y \Leftrightarrow \rho(x, x) = \rho(x, y) = \rho(y, y)$;
   (ii) $\rho(x, x) \leq \rho(x, y)$;
   (iii) $\rho(x, y) = \rho(y, x)$;
   (iv) $\rho(x, z) \leq \rho(x, y) + \rho(y, z) - \rho(y, y)$.

How does a partial metric differ from a metric? Suppose $\rho(x, y) = 0$. Then by (ii) and (iii), $\rho(x, x) = \rho(y, y) = \rho(x, y) = 0$, so by (i) $x = y$. Notice, however, that in general it is **not** the case that if $x = y$, then $\rho(x, y) = 0$. On the other hand, if one assumes that $\rho(x, x) = 0$ for each $x \in X$, then the space (X, ρ) is a metric space in the usual sense.

Some other pertinent facts about partial metrics (see, e.g., [149, 163, 164] for details):

1. Each partial metric ρ on X induces a T_0 topology $\mathcal{T}(\rho)$ on X which has as a base the family of open balls $\{U_\rho(x; \varepsilon) : x \in X, \varepsilon > 0\}$, where $U_\rho(x; \varepsilon) = \{y \in X : \rho(x, y) < \rho(x, x) + \varepsilon\}$. (Originally, the definition was taken as $U_\rho(x; \varepsilon) = \{y \in X : \rho(x, y) < \varepsilon\}$. With this definition, open balls could be empty. Notice that by the definition given here it is always the case that $x \in U_\rho(x; \varepsilon)$ for each $\varepsilon > 0$.) Also, given any two distinct points in X, there is

© Springer International Publishing Switzerland 2014
W. Kirk, N. Shahzad, *Fixed Point Theory in Distance Spaces*,
DOI 10.1007/978-3-319-10927-5_14

a ball containing one that does not contain the other. To see this, suppose $x, y \in X$ with $x \neq y$. Then by (i) and (ii), either $\rho(x, x) < \rho(x, y)$ or $\rho(y, y) < \rho(x, y)$. Suppose $\rho(x, x) < \rho(x, y)$. Let $\varepsilon = \dfrac{\rho(x, y) - \rho(x, x)}{2}$. Then $\rho(x, y) > \varepsilon + \rho(x, x)$, so $x \in U_\rho(x; \varepsilon)$ and $y \notin U_\rho(x; \varepsilon)$. Observe that a sequence $\{x_n\}$ in a partial metric space (X, ρ) *converges* to a point $x \in X$ with respect to $\mathcal{T}(\rho)$ if and only if $\rho(x, x) = \lim_{n \to \infty} \rho(x, x_n)$; in symbols we may write $x_n \overset{\mathcal{T}(\rho)}{\to} x$.

2. If ρ is a partial metric on X, then the function $\rho^s : X \times X \to \mathbb{R}^+$ given by

$$\rho^s(x, y) = 2\rho(x, y) - \rho(x, x) - \rho(y, y), \qquad x, y \in X,$$

is a metric on X. First note that $\rho^s(x, y) \geq 0$ for all $x, y \in X$ by (ii). Also, observe that for $x, y, z \in X$:

   (a) $\rho^s(x, y) = 0 \Leftrightarrow x = y$. To see this, suppose $x = y$. Then $\rho^s(x, x) = 2\rho(x, x) - \rho(x, x) - \rho(x, x) = 0$. Now suppose $\rho^s(x, y) = 0$. Then $2\rho(x, y) - \rho(x, x) - \rho(y, y) = 0$. If $\rho(x, x) = \rho(y, y)$, then $\rho(x, x) = \rho(x, y) = \rho(y, y)$ and by (i) $x = y$. Suppose $\rho(y, y) < \rho(x, x)$. Then $\rho(x, y) < \rho(x, x)$, contradicting (ii).

   (b) $\rho^s(x, y) = 2\rho(x, y) - \rho(x, x) - \rho(y, y)$ and $\rho^s(y, x) = 2\rho(y, x) - \rho(y, y) - \rho(x, x)$, so $\rho^s(x, y) = \rho^s(y, x)$ by (iii).

   (c) $\rho^s(x, z) \leq \rho^s(x, y) + \rho^s(y, z) \Leftrightarrow$
   $2\rho(x, z) - \rho(x, x) - \rho(z, z)$
   $\leq [2\rho(x, y) - \rho(x, x) - \rho(y, y)] + [2\rho(y, z) - \rho(y, y)$
   $-\rho(z, z)] \Leftrightarrow$
   $2\rho(x, z) \leq 2\rho(x, y) + 2\rho(y, z) - 2\rho(y, y)$, which is precisely (iv).

3. A sequence $\{x_n\}$ in (X, ρ) is a *Cauchy sequence* if $\lim_{n,m \to \infty} \rho(x_n, x_m)$ exists and is finite.

4. **Remark.** A sequence $\{x_n\}$ is a Cauchy sequence in (X, ρ) if and only if it is a Cauchy sequence in the metric space (X, ρ^s).

5. A partial metric space (x, ρ) is said to be *complete* if every Cauchy sequence $\{x_n\}$ in X converges, with respect to $\mathcal{T}(\rho)$, to a point $x \in X$ for which $\lim_{n,m \to \infty} \rho(x_n, x_m) = \rho(x, x)$.

6. **Remark.** It is well known that a partial metric space (X, ρ) is complete if and only the metric space (X, ρ^s) is complete. Moreover, given a sequence $\{x_n\}$ in (X, ρ) and $x \in X$ one has $\lim_{n \to \infty} \rho^s(x, x_n) = 0$ if and only if $\rho(x, x) = \lim_{n \to \infty} \rho(x, x_n) = \lim_{n,m \to \infty} \rho(x_n, x_m)$.

7. As mentioned in [200], the success of partial metrics in Computer Science lies in the fact that every partial metric ρ induces a partial

order $\leq_\rho$ on X $(x \leq_\rho y \Leftrightarrow \rho(x,y) = \rho(x,x))$ in such a way that increasing sequences of elements have supremum with respect to $\leq_\rho$ and converge to it with respect the partial metric topology $\mathcal{T}(\rho)$.

14.2. Some Examples

1. Let S^ω be the set of all infinite sequences in $\mathbb{R}$. For all such sequences $x = \{x_i\}$ and $y = \{y_i\}$, let $d_S(x,y) = 2^{-j}$, where j is the largest number (possibly ∞) such that $x_i = y_i$ for all $i < j$. It can be shown that (S^ω, d_S) is a metric space. Now add to S^ω the set S^* of all finite sequences. Then $(S^\omega \cup S^*, d_S)$ is a partial metric space, and $d_S(x,x) \neq 0$ if x is a finite sequence, while $d_S(x,x) = 0$ if x is an infinite sequence.

2. Let $\rho(a,b) = \max\{a,b\}$ for $a,b \in \mathbb{R}^+$. Then $(\mathbb{R}^+, \rho)$ is a partial metric space.

3. Let $\mathcal{I}$ be the collection of nonempty closed bounded intervals in $\mathbb{R}$. For $[a,b], [c,d] \in \mathcal{I}$, let $\rho([a,b], [c,d]) = \max\{b,d\} - \min\{a,c\}$. Then $(\mathcal{I}, \rho)$ is a partial metric space.

14.3. The Partial Metric Contraction Mapping Theorem

THEOREM 14.1 ([149]). *Let (X, ρ) be a complete partial metric space and suppose for some $k \in [0,1)$, $f : X \to X$ satisfies*

$$\rho(f(x), f(y)) \leq k\rho(x,y) \text{ for all } x, y \in X.$$

Then there exists a unique $x^* \in X$ such that $x^* = f(x^*)$ and $\rho(x^*, x^*) = 0$.

PROOF. Suppose $u \in X$. Then for each $n, j \in \mathbb{N}$,

$$\rho(f^{n+j+1}(u), f^n(u))$$
$$\leq \rho(f^{n+j+1}(u), f^{n+j}(u)) + \rho(f^{n+j}(u), f^n(u)) - \rho(f^{n+j}(u), f^{n+j}(u))$$
$$\leq k^{n+j}\rho(f(u), u) + \rho(f^{n+j}(u), f^n(u)).$$

Thus for each $n, j \in \mathbb{N}$

$$\rho(f^{n+j+1}(u), f^n(u))$$
$$\leq k^{n+j}\rho(f(u), u) + k^{n+j-1}\rho(f(u), u) + \rho(f^{n+j-1}(u), f^n(u))$$
$$\leq \cdots$$
$$\leq (k^{n+j} + k^{n+j-1} + \cdots + k^n)\rho(f(u), u) + \rho(f^n(u), f^n(u))$$
$$\leq (k^{n+j} + k^{n+j-1} + \cdots + k^n)\rho(f(u), u) + k^n\rho(u, u)$$
$$= k^n \left(\frac{1 - k^{j+1}}{1 - k}\right)\rho(f(u), u) + k^n\rho(u, u)$$
$$= k^n \left[\left(\frac{1 - k^{j+1}}{1 - k}\right)\rho(f(u), u) + \rho(u, u)\right]$$
$$\leq k^n \left[\left(\frac{\rho(f(u), u)}{1 - k}\right) + \rho(u, u)\right].$$

Therefore, if $m > n$, we see that

$$\rho\left(f^m\left(u\right), f^n\left(u\right)\right) \leq k^n \left[\left(\frac{\rho\left(f\left(u\right), u\right)}{1-k}\right) + \rho\left(u, u\right)\right].$$

It follows that

$$\lim_{m,n \to \infty} \rho\left(f^m\left(u\right), f^n\left(u\right)\right) = 0.$$

Thus for all $m \in \mathbb{N}$,

$$\rho^s\left(f^m\left(u\right), f^n\left(u\right)\right) \leq 2\rho\left(f^m\left(u\right), f^n\left(u\right)\right) \to 0 \text{ as } n \to \infty.$$

Therefore $\{f^n\left(u\right)\}$ is a Cauchy sequence in (X, ρ^s). Since (X, ρ) is complete, so is (X, ρ^s) and the sequence $\{f^n\left(u\right)\}$ converges to some $x^* \in X$ with respect to the metric ρ^s. Therefore

$$\rho\left(x^*, x^*\right) = \lim_{n \to \infty} \rho\left(f^n\left(u\right), x^*\right) = \lim_{n,m \to \infty} \rho\left(f^n\left(u\right), f^m\left(u\right)\right) = 0.$$

Also

$$\begin{aligned}
\rho\left(f\left(x^*\right), x^*\right) &\leq \rho\left(f\left(x^*\right), f^{n+1}\left(u\right)\right) + \rho\left(f^{n+1}\left(u\right), x^*\right) - \rho\left(f^{n+1}\left(u\right), f^{n+1}\left(u\right)\right) \\
&\leq k\rho\left(x^*, f^n\left(u\right)\right) + \rho\left(f^{n+1}\left(u\right), x^*\right).
\end{aligned}$$

Letting $n \to \infty$ we see that $\rho\left(f\left(x^*\right), x^*\right) = 0$. We now have $\rho\left(x^*, x^*\right) = \rho\left(f\left(x^*\right), x^*\right) = 0$ and by (ii) $\rho\left(f\left(x^*\right), f\left(x^*\right)\right) \leq \rho\left(f\left(x^*\right), x^*\right) = 0$, so by (i) $x^* = f\left(x^*\right)$.

Now suppose there exists $y^* \in X$ such that $y^* = f\left(y^*\right)$. Then

$$\rho\left(x^*, y^*\right) = \rho\left(f\left(x^*\right), f\left(y^*\right)\right) \leq k\rho\left(x^*, y^*\right).$$

Since $k < 1$ we conclude $\rho\left(x^*, y^*\right) = 0$ and so $x^* = y^*$. Thus the fixed point is unique. $\qquad\square$

14.4. Caristi's Theorem in Partial Metric Spaces

In order to give an appropriate notion of a Caristi mapping in the framework of partial metric spaces Romaguera [189] proposes two alternatives. Throughout, (X, ρ) is a partial metric space with associated metric space (X, ρ^s).

(ρ-C) A mapping $f : X \to X$ is called a ρ-*Caristi mapping* if there exists a function $\varphi : X \to \mathbb{R}^+$ which is lower semicontinuous for ρ and for which

$$\rho\left(x, f\left(x\right)\right) \leq \varphi\left(x\right) - \varphi\left(f\left(x\right)\right) \text{ for all } x \in X.$$

(ρ^s-C) A mapping $f : X \to X$ is called a ρ^s-*Caristi mapping* if there exists a function $\varphi : X \to \mathbb{R}^+$ which is lower semicontinuous for ρ^s and for which

$$\rho\left(x, f\left(x\right)\right) \leq \varphi\left(x\right) - \varphi\left(f\left(x\right)\right) \text{ for all } x \in X.$$

It is clear that every ρ-Caristi mapping is a ρ^s-Caristi mapping, but the converse need not be true. [Suppose φ is lower semicontinuous for ρ. This means that $x_n \xrightarrow{\mathcal{T}(\rho)} x$ and $\varphi(x_n) \to r \Rightarrow \varphi(x) \le r$. However $x_n \xrightarrow{\rho^s} x \Rightarrow$ $x_n \xrightarrow{\mathcal{T}(\rho)} x$. To see this, suppose $x_n \xrightarrow{\rho^s} x$ and let $\varepsilon > 0$. Then there exists $n_0 \in \mathbb{N}$ such that

$$n \ge n_0 \Rightarrow \rho^s(x_n, x) = 2\rho(x, x_n) - \rho(x, x) - \rho(x_n, x_n) \le \varepsilon.$$

But by (ii), $\rho(x, x_n) - \rho(x, x) \le 2\rho(x, x_n) - \rho(x, x) - \rho(x_n, x_n)$. Hence $n \ge n_0 \Rightarrow x_n \in U_\rho(x; \varepsilon)$, i.e., $x_n \xrightarrow{\mathcal{T}(\rho)} x$.]

In a first attempt to generalize Kirk's characterization of metric completeness to the partial metric framework, one could conjecture that a partial metric space is complete if and only if every ρ-Caristi mapping has a fixed point. However the following example shows that this conjecture is false.

Example ([189]). On the set $\mathbb{N}$ of natural numbers construct the partial metric ρ given by

$$\rho(n, m) = \max\left\{\frac{1}{n}, \frac{1}{m}\right\}, \qquad n, m \in \mathbb{N}.$$

Properties (i)–(iii) of the definition of a partial metric are trivial. To see that (iv) holds, suppose $n, m, p \in \mathbb{N}$. We need to show that

$$\rho(n, p) \le \rho(n, m) + \rho(m, p) - \rho(m, m).$$

However $\max\left\{\frac{1}{n}, \frac{1}{p}\right\} \le \max\left\{\frac{1}{n}, \frac{1}{m}\right\} + \max\left\{\frac{1}{m}, \frac{1}{p}\right\}$. Case (1). Suppose $n \le p$. Then the left side is $\frac{1}{n}$, while $\max\left\{\frac{1}{n}, \frac{1}{m}\right\} \ge \frac{1}{n}$ and $\max\left\{\frac{1}{m}, \frac{1}{p}\right\} - \frac{1}{m} \ge 0$. Case (2). Suppose $p \le n$. Then the left side is $\frac{1}{p}$ while $\max\left\{\frac{1}{m}, \frac{1}{p}\right\} \ge \frac{1}{p}$ and $\max\left\{\frac{1}{n}, \frac{1}{m}\right\} - \frac{1}{m} \ge 0$. Thus (iv) holds.

Notice that if $m > n$,

$$
\begin{aligned}
\rho^s(n, m) &= 2\rho(n, m) - \rho(n, n) - \rho(m, m) \\
&= 2\max\left\{\frac{1}{n}, \frac{1}{m}\right\} - \frac{1}{n} - \frac{1}{m} \\
&= \frac{2}{n} - \frac{1}{n} - \frac{1}{m} \\
&= \frac{1}{n} - \frac{1}{m}.
\end{aligned}
$$

Thus $\{n\}_{n \in \mathbb{N}}$ is a Cauchy sequence in $(\mathbb{N}, \rho^s)$. However $\{n\}_{n \in \mathbb{N}}$ does not have a limit, and so $(\mathbb{N}, \rho^s)$, hence $(\mathbb{N}, \rho)$, is not complete. Also there are no fixed-point free ρ-Caristi mappings defined on $\mathbb{N}$.

Indeed, let $f : \mathbb{N} \to \mathbb{N}$ and suppose there is a lower semicontinuous mapping φ from $(\mathbb{N}, \mathcal{T}(\rho))$ into $\mathbb{R}^+$ such that

$$\rho(n, f(n)) \leq \varphi(n) - \varphi(f(n)).$$

If $1 < f(1)$, $\rho(1, f(1)) = 1 = \rho(1,1)$. This means that $f(1) \in U_\rho(1; \varepsilon)$ for any $\varepsilon > 0$, so $\varphi(1) \leq \varphi(f(1))$ by lower semicontinuity of φ. (It is possible to think of it this way. The set $\{x \in \mathbb{N} : \varphi(x) \leq \varphi(f(1))\}$ is closed. Obviously 1 is in the closure of this set. Another point of view: Define $\{x_n\}$ by setting $x_n \equiv f(1)$. Then $f(1) \in U_\rho(1; \varepsilon)$ for any $\varepsilon > 0$ means $x_n \overset{\mathcal{T}(\rho)}{\to} 1$. On the other hand, $\varphi(x_n) \equiv \varphi(f(1)) \to \varphi(f(1))$.) Since $\rho(1, f(1)) \leq \varphi(1) - \varphi(f(1))$, we conclude $\rho(1, f(1)) = 0$, which contradicts $\rho(1, f(1)) = 1$. We conclude therefore that there does not exist a fixed-point free ρ-Caristi mapping from $\mathbb{N} \to \mathbb{N}$.

DEFINITION 14.1 ([189]). A sequence $\{x_n\}_{n \in \mathbb{N}}$ in a partial metric space (X, ρ) is called 0-*Cauchy* if $\lim_{n,m \to \infty} \rho(x_n, x_m) = 0$. The space (X, ρ) is said to be 0-*complete* if every 0-Cauchy sequence in X converges, with respect to $\mathcal{T}(\rho)$, to a point $x \in X$ for which $\rho(x, x) = 0$. Of course, every complete partial metric space is 0-complete but the converse is not true in general.

It is known that the fixed point property for Caristi maps characterizes metric completeness. Specifically, if (X, d) is a metric space, a mapping $f : X \to X$ is said to be a *Caristi mapping* if there exists a lower semicontinuous mapping $\varphi : X \to \mathbb{R}$ which is bounded below and for which

$$d(x, f(x)) \leq \varphi(x) - \varphi(f(x))$$

for every $x \in X$.

THEOREM 14.2 ([113]). *A metric space (X, d) is complete if and only if every Caristi mapping $f : X \to X$ has a fixed point.*

Subsequently there have been several similar characterizations of completeness (see, e.g., [215, 207], and papers cited in [189]). In particular, it is known [30] that a normed linear space is complete if and only if every contraction mapping defined on the space has a fixed point.

We now discuss Romaguera's paper [189] in which an analog of Theorem 14.2 is given for partial metric spaces.

LEMMA 14.1. *Let (X, ρ) be a partial metric space. Then for each $x \in X$ the function $\rho_x : X \to \mathbb{R}^+$ defined by $\rho_x(y) = \rho(x, y)$ is lower semicontinuous for (X, ρ^s).*

PROOF. Assume that $\lim_{n\to\infty} \rho^s(y_n, y) = 0$. Then, since $\rho^s(y_n, y) = 2\rho(y_n, y) - \rho(y_n, y_n) - \rho(y, y)$ and $\rho(y, y) \leq \rho(y, y_n)$,

$$
\begin{aligned}
\rho_x(y) &\leq \rho_x(y_n) + \rho(y_n, y) - \rho(y_n, y_n) \\
&= \rho_x(y_n) + \rho^s(y_n, y) - \rho(y_n, y) + \rho(y, y) \\
&\leq \rho_x(y_n) + \rho^s(y_n, y).
\end{aligned}
$$

Since $\rho(y, y) \leq \rho(y, y_n)$ we have $\rho_x(y) \leq \liminf_{n\to\infty} \rho_x(y_n)$.   □

The main result of [189] is the following:

THEOREM 14.3. *A partial metric space (X, ρ) is 0-complete if and only if every ρ^s-Caristi mapping has a fixed point.*

PROOF. ([189]) Suppose that (X, ρ) is 0-complete, and let f be a ρ^s-Caristi mapping on X. Then there exists a ρ^s-lower semicontinuous function $\varphi : X \to \mathbb{R}^+$ for which

$$
\rho(x, f(x)) \leq \varphi(x) - \varphi(f(x)) \text{ for all } x \in X.
$$

For each $x \in X$ set

$$
A_x := \{y \in X : \rho(x, y) \leq \phi(x) - \phi(y)\}.
$$

$f(x) \in A_x \Rightarrow A_x \neq \emptyset$. Moreover A_x is closed in the metric space (X, ρ^s) since the mapping $y \mapsto \rho(x, y) + \phi(y)$ is lower semicontinuous in (X, ρ^s).

Now fix $x_0 \in X$ and choose $x_1 \in A_{x_0}$ so that $\phi(x_1) < \inf_{y \in A_{x_0}} \phi(y) + 2^{-1}$. Clearly $A_{x_1} \subseteq A_{x_0}$. Hence for each $x \in A_{x_1}$ we have

$$
\begin{aligned}
\rho(x_1, x) &\leq \phi(x_1) - \phi(x) \leq \inf_{y \in A_{x_0}} \phi(y) + 2^{-1} - \phi(x) \\
&\leq \phi(x) + 2^{-1} - \phi(x) = 2^{-1}.
\end{aligned}
$$

Continuing in this manner it is possible to construct a sequence $\{x_n\}$ in X such that the associated $\{A_{x_n}\}$ of closed subsets of (X, ρ^s) satisfies

(i) $A_{x_{n+1}} \subseteq A_{x_n}$ and $x_{n+1} \in A_{x_n}$ for all $n \in \mathbb{N}$;

(ii) $\rho(x_n, x) \leq 2^{-n}$ for all $x \in A_{x_n}$, $n \in \mathbb{N}$.

Since $\rho(x_n, x_n) \leq \rho(x_n, x_{n+1})$ and, by (i) and (ii), $\rho(x_n, x_m) \leq 2^{-n}$ for all $m > n$, it follows that $\lim_{n,m\to\infty} \rho(x_n, x_m) = 0$. Therefore $\{x_n\}$ is a 0-Cauchy sequence in (X, ρ). Since (X, ρ) is 0-complete, there exists $x^* \in X$ such that $\lim_{n\to\infty} \rho(x^*, x_n) = \rho(x^*, x^*) = 0$, and thus $\lim_{n\to\infty} \rho^s(x^*, x_n) = 0$. Therefore $x^* \in \cap_{n\in\mathbb{N}} A_{x_n}$.

To see that $x^* = f(x^*)$, note that

$$
\begin{aligned}
\rho(x_n, f(x^*)) &\leq \rho(x_n, x^*) + \rho(x^*, f(x^*)) \\
&\leq \phi(x_n) - \phi(x^*) + \phi(x^*) - \phi(f(x^*)),
\end{aligned}
$$

for all $n \in \mathbb{N}$. Consequently $f(x^*) \in \cap_{n\in\mathbb{N}} A_{x_n}$, so by (ii) $\rho(x_n, f(x^*)) < 2^{-n}$ for all $n \in \mathbb{N}$. Since $\rho(x^*, f(x^*)) \leq \rho(x^*, x_n) + \rho(x_n, f(x^*))$ and

$\lim_{n \to \infty} \rho(x^*, x_n) = 0$, it follows that $\rho(x^*, f(x^*)) = 0$. Hence $\rho^s(x^*, f(x^*)) = 0$, and since $\rho^s(x^*, f(x^*)) \le 2\rho(x^*, f(x^*))$ we conclude that $f(x^*) = x^*$.

For the converse, suppose that there is a 0-Cauchy sequence $\{x_n\}$ of distinct points in (X, ρ) which is not convergent in (X, ρ^s). Select a subsequence $\{y_n\}$ if $\{x_n\}$ for which $\rho(y_n, y_{n+1}) < 2^{-n}$ for all $n \in \mathbb{N}$. Let $A := \{y_n : n \in \mathbb{N}\}$, and define $f : X \to X$ by setting $f(x) = y_0$ if $x \in X \backslash A$ and $f(y_n) = y_{n+1}$ for all $n \in \mathbb{N}$. Also note that A is closed in (X, ρ^s).

Now define $\phi : X \to \mathbb{R}^+$ by setting $\phi(x) = \rho(x, y_0) + 1$ if $x \in X \backslash A$ and $\phi(y_n) = 2^{-n}$ for all $n \in \mathbb{N}$. Then $\phi(y_{n+1}) < \phi(y_n)$ for all $n \in \mathbb{N}$, and $\phi(y_0) \le \phi(x)$ for all $x \in X \backslash A$. From this fact and Lemma 14.1 we conclude that ϕ is lower semicontinuous on (X, ρ^s). Moreover, for each $x \in X \backslash A$,

$$\rho(x, f(x)) = \rho(x, y_0) = \phi(x) - \phi(y_0) = \phi(x) - \phi(f(x)),$$

and for each $y_n \in A$,

$$\begin{aligned} \rho(y_n, f(y_n)) &= \rho(y_n, y_{n+1}) < 2^{-(n+1)} \\ &= \phi(y_n) - \phi(y_{n+1}) \\ &= \phi(y_n) - \phi(f(y_n)). \end{aligned}$$

Therefore f is a Caristi ρ^s-mapping which has no fixed point.   □

14.5. Nadler's Theorem in Partial Metric Spaces

Following [13] we introduce the notion of a partial Hausdorff metric. Let (X, ρ) be a partial metric space and let $\mathcal{CB}^\rho(X)$ be the family of all nonempty bounded $\mathcal{T}(\rho)$-closed subsets of (X, ρ). (Here a set A in X is bounded if there exists $x_0 \notin X$ and $M \ge 0$ such that for all $a \in A$, $\rho(x_0, a) \le \rho(a, a) + M$.)

Now for $A, B \in \mathcal{CB}^\rho(X)$ and $x \in X$, set

$$\begin{aligned} dist_\rho(x, A) &= \inf\{\rho(x, a) : a \in A\}; \\ \delta_\rho(A, B) &= \sup\{dist_\rho(a, B) : a \in A\}; \\ \delta_\rho(B, A) &= \sup\{dist_\rho(b, A) : b \in B\}. \end{aligned}$$

Finally, let $H_\rho(A, B) = \max\{\delta_\rho(A, B), \delta_\rho(B, A)\}$. It is easy to check that

$$dist_\rho(x, A) = 0 \Rightarrow dist_{\rho^s}(x, A) = 0,$$

where $dist_{\rho^s}(x, A) = \inf\{\rho^s(x, a) : a \in A\}$.

The following is the main result of [13].

THEOREM 14.4. Let (X, ρ) be a complete partial metric space, and suppose $T : X \to \mathcal{CB}^\rho(X)$ satisfies for some $k \in (0, 1)$ and all $x, y \in X$:

$$H_\rho(T(x), T(y)) \le k\rho(x, y).$$

Then T has a fixed point (i.e., for some $x^* \in X$, $x^* \in T(x^*)$).

We should remark that this theorem, while true, requires the assumption that $\mathcal{CB}^\rho(X) \neq \emptyset$ to be meaningful. In [**190**] Romaguera gives an example of a complete partial metric space (X, ρ) for which $\mathcal{CB}^\rho(X) = \emptyset$.

Before proving the theorem, we make some preliminary observations, which are taken from [**13**]. Throughout (X, ρ) denotes a partial metric space.

PROPOSITION 14.1. *For $A, B, C \in \mathcal{CB}^\rho(X)$:*

(i) $\delta_\rho(A, A) = \sup\{\rho(a, a) : a \in A\}$;

(ii) $\delta_\rho(A, A) \leq \delta_\rho(A, B)$;

(iii) $\delta_\rho(A, B) = 0 \Rightarrow A \subseteq B$;

(iv) $\delta_\rho(A, B) \leq \delta_\rho(A, C) + \delta_\rho(C, B) - \inf_{c \in C}\{\rho(c, c)\}$.

PROOF. (i) Let $A \in \mathcal{CB}^\rho(X)$. Since a is in the closure of A if and only if $dist_\rho(a, A) = \rho(a, a)$,

$$\delta_\rho(A, A) = \sup\{dist_\rho(a, A) : a \in A\} = \sup\{\rho(a, a) : a \in A\}.$$

(ii) Let $a \in A$. Since $\rho(a, a) \leq \rho(a, b)$ for all $b \in B$, it follows that $\rho(a, a) \leq dist_\rho(a, B) \leq \delta_\rho(A, B)$. From (i) we conclude that

$$\delta_\rho(A, A) = \sup\{\rho(a, a) : a \in A\} \leq \delta_\rho(A, B).$$

(iii) Suppose that $\delta_\rho(A, B) = 0$. Then $dist_\rho(a, B) = 0$ for all $a \in A$. From (i) and (ii) it follows that $\rho(a, a) \leq \delta_\rho(A, B) = 0$ for all $a \in A$. Hence $dist_\rho(a, B) = \rho(a, a)$ for all $a \in A$. Thus a is in the closure of B for all $a \in A$ and, since B is closed, $A \subseteq B$.

(iv) Let $a \in A$, $b \in B$, $c \in C$. Since

$$\rho(a, b) \leq \rho(a, c) + \rho(c, b) - \rho(c, c),$$

it follows that

$$dist_\rho(a, B) \leq \rho(a, c) + dist_\rho(c, B) - \rho(c, c);$$

whence

$$dist_\rho(a, B) + \rho(c, c) \leq \rho(a, c) + \delta_\rho(C, B).$$

Since $c \in C$ is arbitrary,

$$dist_\rho(a, B) + \inf_{c \in C} \rho(c, c) \leq dist_\rho(a, C) + \delta_\rho(C, B).$$

Since $a \in A$ is arbitrary, it follows that

$$\delta_\rho(A, B) \leq \delta_\rho(A, C) + \delta_\rho(C, B) - \inf_{c \in C} \rho(c, c).$$

$\square$

PROPOSITION 14.2. *For all $A, B, C \in \mathcal{CB}^\rho(X)$*

(H1) $H_\rho(A, A) \leq H_\rho(A, B)$;

(H2) $H_\rho(A, B) = H_\rho(B, A)$;

(H3) $H_\rho(A, B) \leq H_\rho(A, C) + H_\rho(C, B) - \inf_{c \in C} \rho(c, c)$.

PROOF. From (ii) of Proposition 14.1 we have

$$H_\rho(A, A) = \delta_\rho(A, A) \leq \delta_\rho(A, B) \leq H_\rho(A, B).$$

$(H2)$ is obvious from the definition of H_ρ. Using (iv) of Proposition 14.1 we have

$$
\begin{aligned}
H_\rho(A, B) &= \max\{\delta_\rho(A, B), \delta_\rho(B, A)\} \\
&\leq \max\left\{\begin{array}{l} \delta_\rho(A, C) + \delta_\rho(C, B) - \inf_{c \in C} \rho(c, c), \\ \delta_\rho(B, C) + \delta_\rho(C, A) - \inf_{c \in C} \rho(c, c) \end{array}\right\} \\
&= \max\{\delta_\rho(A, C) + \delta_\rho(C, B), \delta_\rho(B, C) + \delta_\rho(C, A)\} \\
&\quad - \inf_{c \in C} \rho(c, c) \\
&\leq \max\{\delta_\rho(A, C), \delta_\rho(C, A)\} + \max\{\delta_\rho(C, B), \delta_\rho(B, C)\} \\
&\quad - \inf_{c \in C} \rho(c, c) \\
&= H_\rho(A, C) + H_\rho(C, B) - \inf_{c \in C} \rho(c, c).
\end{aligned}
$$

$\square$

COROLLARY 14.1. For $A, B \in \mathcal{CB}^\rho(X)$,

$$H_\rho(A, B) = 0 \Rightarrow A = B.$$

PROOF. Suppose $H_\rho(A, B) = 0$. Then by definition, $\delta_\rho(A, B) = \delta_\rho(B, A) = 0$. By (iii) of Proposition 14.1, $A \subseteq B$ and $B \subseteq A$; hence $A = B$. $\square$

REMARK 14.1. *An example given in [13] shows that in general the converse of the above corollary is not true.*

LEMMA 14.2. *Let* $A, B \in \mathcal{CB}^\rho(X)$. *Then for any* $h > 1$ *and* $a \in A$ *there exists* $b = b(a) \in B$ *such that*

(14.1) $$\rho(a, b) \leq hH_\rho(A, B).$$

PROOF. Suppose $A = B$. Then by (i) of Proposition 14.1

$$H_\rho(A, B) = H_\rho(A, A) = \delta_\rho(A, A) = \sup_{x \in A} \rho(x, x).$$

However since $h > 1$, if $a \in A$

$$\rho(a, a) \leq \sup_{x \in A} \rho(x, x) = H_\rho(A, B) \leq hH_\rho(A, B).$$

Consequently $b = a$ satisfies (14.1).

Now suppose $A \neq B$, and suppose there exists $a \in A$ such that $\rho(a, b) > hH_\rho(A, B)$ for all $b \in B$. This implies that $\inf\{\rho(a, y) : y \in B\} \geq hH_\rho(A, B)$. Thus $dist_\rho(a, B) \geq hH_\rho(A, B)$. However

$$H_\rho(A, B) \geq \delta_\rho(A, B) = \sup_{x \in A} dist_\rho(x, B) \geq dist_\rho(a, B) \geq hH_\rho(A, B).$$

Since $A \neq B$, Corollary 14.1 implies $H_\rho(A, B) > 0$, and this is a contradiction. $\square$

PROOF OF THEOREM 14.4. ([13]) Let $x_0 \in X$ and $x_1 \in T(x_0)$. From Lemma 14.2 there exists $x_2 \in T(x_1)$ such that $\rho(x_1, x_2) \leq \frac{1}{\sqrt{k}} H_\rho(T(x_0), T(x_1))$. Since

$$H_\rho(T(x_0), T(x_1)) \leq k\rho(x_0, x_1),$$

it follows that $\rho(x_1, x_2) \leq \sqrt{k}\rho(x_0, x_1)$. Similarly there exists $x_3 \in T(x_2)$ such that

$$\rho(x_2, x_3) \leq \frac{1}{\sqrt{k}} H_\rho(T(x_1), T(x_2)) \leq \sqrt{k}\rho(x_1, x_2).$$

Inductively, there exists a sequence $\{x_n\} \subset X$ such that

$$x_{n+1} \in T(x_n) \text{ and } \rho(x_{n+1}, x_n) \leq \sqrt{k}\rho(x_n, x_{n-1}) \text{ for all } n \geq 1.$$

By (iv) of the definition of a partial metric, for any $n, m \in \mathbb{N}$,

$$
\begin{aligned}
\rho(x_n, x_{n+m}) &\leq \rho(x_n, x_{n+1}) + \rho(x_{n+1}, x_{n+2}) + \cdots + \rho(x_{n+m-1}, x_{n+m}) \\
&\leq \left(\sqrt{k}\right)^n \rho(x_0, x_1) + \left(\sqrt{k}\right)^{n-1} \rho(x_0, x_1) + \\
&\quad \cdots + \left(\sqrt{k}\right)^{n+m-1} \rho(x_0, x_1) \\
&= \left(\left(\sqrt{k}\right)^n + \left(\sqrt{k}\right)^{n+1} + \cdots + \left(\sqrt{k}\right)^{n+m-1}\right) \rho(x_0, x_1) \\
&\leq \frac{\left(\sqrt{k}\right)^n}{1 - \sqrt{k}} \rho(x_0, x_1) \to 0 \text{ as } n \to \infty.
\end{aligned}
$$

Thus for all $m \in \mathbb{N}$,

$$\rho^s(x_n, x_{n+m}) \leq 2\rho(x_n, x_{n+m}) \to 0 \text{ as } n \to \infty.$$

Therefore $\{x_n\}$ is a Cauchy sequence in (X, ρ^s). Since (X, ρ) is complete, as noted earlier so is (X, ρ^s) and the sequence $\{x_n\}$ converges to some $x^* \in X$ with respect to the metric ρ^s. Therefore

$$(14.2) \qquad \rho(x^*, x^*) = \lim_{n \to \infty} \rho(x_n, x^*) = \lim_{n,m \to \infty} \rho(x_n, x_m) = 0.$$

Since $H_\rho(T(x_n), T(x^*)) \leq k\rho(x_n, x^*)$ it follows that $\lim_{n \to \infty} H_\rho(T(x_n), T(x^*)) = 0$. However $x_{n+1} \in T(x_n)$; hence

$$dist_\rho(x_{n+1}, T(x^*)) \leq \delta_\rho(T(x_n), T(x^*)) \leq H_\rho(T(x_n), T(x^*))$$

and it follows that $\lim_{n \to \infty} dist_\rho(x_{n+1}, T(x^*)) = 0$. On the other hand

$$dist_\rho(x^*, T(x^*)) \leq \rho(x^*, x_{n+1}) + dist_\rho(x_{n+1}, T(x^*)).$$

Letting $n \to \infty$ and using (14.2) it follows that $dist_\rho(x^*, T(x^*)) = 0$. Therefore $\rho(x^*, x^*) = dist_\rho(x^*, T(x^*))$ and since $T(x^*)$ is closed, $x^* \in T(x^*)$. □

14.6. Further Remarks

Recently it has been shown that many fixed point results proved in the context to partial metric framework can be obtained from their corresponding metric counterparts (see [**90**, **95**]). Specifically, it was proved in [**95**] that every partial metric ρ on a nonempty set X induces a metric d_ρ on X such that $\mathcal{T}(\rho^s) \subseteq \mathcal{T}(d_\rho)$, where

$$d_\rho(x, y) = \begin{cases} 0 & \text{if } x = y \\ \rho(x, y) & \text{if } x \neq y \end{cases}.$$

Moreover, (X, d_ρ) is complete if and only if (X, ρ) is 0-complete.

Taking these facts into account, it was pointed out in [**90**] that a wide class of generalized contractive mappings in the partial metric context are at the same time generalized contractive mappings in the metric setting as well, so the existence and uniqueness of fixed point results for such mappings can be deduced from those results given for the same kind of mappings in the metric case. As a particular case of this observation one sees that if a mapping $f : X \to X$ satisfies the contractive condition

(14.3) $\rho(f(x), f(y)) \leq k\rho(x, y)$

for all $x, y \in X$ and some $k \in [0, 1)$, then

$$d_\rho(f(x), f(y)) \leq k d_\rho(x, y)$$

for all $x, y \in X$. Therefore if the partial metric space (X, ρ) is 0-complete and if $f : X \to X$ satisfies (14.3) the existence and uniqueness of a fixed point of f follows from the classical Banach contraction mapping theorem. However, as noted in [**200**], Theorem 14.1 provides a property of such a fixed point that cannot be deduced in the metric context and that is essential in Denotational Semantics. Specifically, if x^* is the fixed point of the mapping f, then $\rho(x^*, x^*) = 0$. As a result, the classical fixed point results do not invalidate totally the new ones in the partial metric framework. As noted in [**200**], this is especially true of the following result.

THEOREM 14.5 ([**200**]). *Let (X, ρ) be a 0-complete partial metric space, let $f : X \to X$ be monotone relative to the partial order $\leq_\rho$, and suppose $x_0 \in X$ satisfies $x_0 \leq f(x_0)$. If there exists $k \in [0, 1)$ such that*

$$\rho(f^n(x_0), f^n(x_0)) \leq k\rho(f^{n-1}(x_0), f^{n-1}(x_0))$$

for all $n \in \mathbb{N}$, then f has a fixed point $x^* \in X$ such that

1) *x^* is the unique fixed point of f in $\{z \in X : x_0 \leq_\rho z\}$.*
2) *x^* is the supremum of $\{f^n(x_0)\}_{n \in \mathbb{N}}$ in $(X, \leq_\rho)$ and maximal in $(X, \leq_\rho)$.*
3) *The sequence $\{f^n(x_0)\}_{n \in \mathbb{N}}$ converges to x^* with respect to $\mathcal{T}(\rho^s)$.*
4) *$\rho(x^*, x^*) = 0$.*

CHAPTER 15

Diversities

15.1. Introduction

A generalization of metric spaces called "diversities" has been introduced by Bryant and Tupper in [**43**]. It is shown there that remarkable analogies exist between hyperconvex metric spaces and diversities, especially involving the "tight span" (otherwise called the injective or hyperconvex envelop).

DEFINITION 15.1. Let X be a set, and let (X) denote the collection of finite subsets of X. A *diversity* is a pair (X, δ), where $\delta : (X) \to \mathbb{R}$ satisfies for all $A, B, C \in (X)$:

(D1) $\delta(A) \geq 0$, and $\delta(A) = 0$ if and only if $|A| \leq 1$;
(D2) if $B \neq \emptyset$, $\delta(A \cup C) \leq \delta(A \cup B) + \delta(B \cup C)$.

A diversity (X, δ) is said to be *bounded* if there exist $M \in \mathbb{R}$ such that $\delta(A) \leq M$ for all $A \in (X)$.

Motivation for the use of the term diversity stems from the appearance of special cases of the definition in work on phylogenetic and ecological diversities [**80, 155, 168, 205**].

The following are among examples of diversities given in [**43**].

1. *Diameter diversity*: Let (X, d) be a metric space. For all $A \in (X)$, let

$$\delta(A) = \max\{d(a, b) : a, b \in A\} = diam(A).$$

   Then (X, δ) is a diversity called the *diameter diversity*.
2. *Phylogenetic diversity*: Let (T, d) be an $\mathbb{R}$-tree and let μ be the one-dimensional Hausdorff measure on it. In this case, $\mu([a, b]) = d(a, b)$ for $a, b \in T$. If $A \subseteq T$,

$$conv(A) = \bigcup_{a,b \in A} [a, b].$$

   For $A \in (T)$, set

$$\delta_t(A) = \mu(conv(A)).$$

© Springer International Publishing Switzerland 2014
W. Kirk, N. Shahzad, *Fixed Point Theory in Distance Spaces*,
DOI 10.1007/978-3-319-10927-5_15

Then δ_t defines a diversity on T called the $\mathbb{R}$-*tree* (or *real tree*) *diversity* on T. Finally, a diversity (X, δ) is called a *phylogenetic diversity* if it can be embedded in an $\mathbb{R}$-tree diversity for some complete $\mathbb{R}$-tree (T, d).

Following [**43**] we now list some basic properties of diversities.

PROPOSITION 15.1. *Let (X, δ) be a diversity. Then:*

1. δ *is monotone, that is,* $\delta(A) \leq \delta(B)$ *for $A, B \in (X)$ with $A \subseteq B$.*
2. δ *induces a metric* $d : X \times X \to \mathbb{R}$ *on X defined by* $d(x, y) = \delta(\{x, y\})$ *for $x, y \in X$.*
3. *For $A, B \in (X)$, if $A \cap B \neq \emptyset$, then* $\delta(A \cup B) \leq \delta(A) + \delta(B)$.

PROOF.   1. For any $A \in (X)$ and $b \in X$, then by $(D2)$ (taking $C = \emptyset$) we have

$$\delta(A) \leq \delta(A \cup \{b\}) + \delta\{b\} = \delta(A \cup \{b\}).$$

The result now follows by induction. Let $A \in (X)$, $n > 1$, and $\{b_1, \cdots, b_n\} \in (X)$, and assume

$$\delta(A) \leq \delta(A \cup \{b_1, \cdots, b_{n-1}\}).$$

Again by $(D2)$

$$\begin{aligned}
\delta(A) &\leq \delta(A \cup \{b_1, \cdots, b_{n-1}\}) \\
&\leq \delta(A \cup \{b_1, \cdots, b_{n-1}\} \cup \{b_n\}) \\
&= \delta(A \cup \{b_1, \cdots, b_n\}).
\end{aligned}$$

2. We have $d(x, y) = 0$ if and only if $x = y$ by $(D1)$. Symmetry of d is clear. The triangle inequality follows from $(D2)$:

$$d(x, z) = \delta(\{x, z\}) \leq \delta(\{x, y\}) + \delta(\{y, z\}) = d(x, y) + d(y, z)$$

for all $x, y, z \in X$.
3. Since $A \cap B \neq \emptyset$, $(D2)$ implies

$$\delta(A \cup B) \leq \delta(A \cup (A \cap B)) + \delta(B \cup (A \cap B)) = \delta(A) + \delta(B).$$

$\square$

We remark in particular that statements 1 and 2 in the above proposition imply the following: If $A \in (X)$ and $z \in X$, then

$$\delta(A) \leq \delta(A \cup \{z\}) \leq \sum_{a \in A} \delta(\{z, a\}) = \sum_{a \in A} d(z, a).$$

15.2. Hyperconvex Diversities

DEFINITION 15.2.   *1.* Given diversities (Y_1, δ_1) and (Y_2, δ_2), a mapping $f : Y_1 \to Y_2$ is said to be *nonexpansive* if for all $A \subseteq (Y_1)$

$$\delta_2(f(A)) \leq \delta_1(A),$$

and f is said to be an *embedding* if it is one-to-one and $\delta_2(f(A)) = \delta_1(A)$ for all $A \in (Y_1)$.

2. A diversity is *injective* if it satisfies the following property: given any pair of diversities (Y_1, δ_1), (Y_2, δ_2) and embedding $\pi : Y_1 \to Y_2$ and a nonexpansive map $f : Y_1 \to X$ there is a nonexpansive map $g : Y_2 \to X$ such that $f = g \circ \pi$.

3. A diversity (X, δ) is said to be *hyperconvex* if for all $r : (X) \to \mathbb{R}$ such that

$$\delta\left(\bigcup_{A \in \mathcal{A}} A\right) \leq \sum_{A \in \mathcal{A}} r(A)$$

for all finite $\mathcal{A} \subseteq (X)$ there is a $z \in X$ such that $\delta(\{z\} \cup Y) \leq r(Y)$ for all finite $Y \subseteq X$.

It is worth noting that if (X, δ) is a hyperconvex diversity and d is its induced metric, then (X, d) is complete (see Proposition 3.10 of [77]). It is also proved in [77] that (X, d) need not be hyperconvex.

The following is the diversity counterpart of the fundamental result of Aronszajn and Panitchpakdi's result of hyperconvex metric spaces (see Theorem 4.2 of Chap. 4).

THEOREM 15.1. *A diversity is injective if and only if it is hyperconvex.*

15.3. Fixed Point Theory

As we have observed earlier, any bounded hyperconvex metric space has the fixed point property for nonexpansive mappings. However it has been shown in [77] that if (X, δ) is a hyperconvex diversity for which its induced metric space (X, d) is bounded, then (X, d) need **not** have the fixed point property for nonexpansive mappings. However if the diversity δ is bounded, then (X, d) does have the fixed point property for nonexpansive mappings. Indeed, the following is Theorem 4.2 of [77].

THEOREM 15.2. *Let (X, δ) be a bounded and hyperconvex diversity with induced metric space (X, d) and suppose $f : X \to X$ is nonexpansive relative to d. Then f has a fixed point.*

For a bounded subset A of a metric space (X, d) let $r_x(A)$ denote the Chebyshev radius of A relative to $x \in X$, that is

$$r_x(A) = \inf\{r \geq 0 : A \subseteq B(x; r)\}$$

Recall also that

$$cov(A) := \bigcap \{D : D \text{ is a closed ball, and } D \supseteq A\}.$$

For the proof of the theorem we will need the simple fact that $r_x(A) = r_x(cov(A))$

PROOF OF THEOREM 15.2. ([77]) Let

$$\mathcal{U} = \{A \subseteq X : A \neq \emptyset, \ A = cov(A), \ f(A) \subseteq A\}.$$

Our first objective is to show that $\mathcal{U}$ has a minimal element. First, since δ is a bounded diversity, (X, d) is a bounded metric space. Thus $\mathcal{U} \neq \emptyset$ because $X \in \mathcal{U}$. Now let $\{A_i\}_{i \in I}$ be a decreasing chain in $\mathcal{U}$ ordered by set inclusion. We shall show that $\cap_{i \in I} A_i \neq \emptyset$.

Notice that for each admissible subset A of X,

$$A = \bigcap_{x \in X} B(x; r_x(A)).$$

For each $x \in X$ and $i, j \in I$, $A_i \subseteq A_j$ if $i \geq j$; hence $r_x(A_i) \leq r_x(A_j)$. Hence it is possible to define

$$r(x) = \inf \{r_x(A_i) : i \in I\}.$$

Notice that if $r(x) = 0$ for some $x \in X$, then $x \in \cap_{i \in I} A_i$, so we assume $r(x) > 0$ for each x. Let $\{y_1, \cdots, y_n\}$ be a finite collection of points of X and let $\varepsilon > 0$. Then for each $k \in \{1, \cdots, n\}$ there exist $i(k)$ such that

$$r_{y_k}(A_{i(k)}) \leq r(y_k) + \varepsilon.$$

We may further assume that $A_{i(1)} \subseteq A_{i(2)} \subseteq \cdots \subseteq A_{i(n)}$. Hence

$$r_{y_k}(A_{i(1)}) \leq r(y_k) + \varepsilon.$$

Taking any $a \in A_{i(1)}$ we have $d(y_k, a) \leq r_{y_k}(A_{i(1)}) \leq r(y_k) + \varepsilon$, and thus

$$\delta(\{y_1, \cdots, y_n\}) \leq \sum_{k=1}^{n} \delta(\{y_k, a\}) = \sum_{k=1}^{n} d(y_k, a) \leq \sum_{k=1}^{n} r(y_k) + n\varepsilon.$$

Since $\varepsilon > 0$ is arbitrary,

$$\delta(\{y_1, \cdots, y_n\}) \leq \sum_{k=1}^{n} r(y_k)$$

for any finite collection $\{y_1, \cdots, y_n\}$ of point in X. Therefore for a given finite collection $\{x_1, \cdots, x_m\} \subseteq X$ we can set

$$r(\{x_1, \cdots, x_m\}) = \sum_{i=1}^{m} r(x_k).$$

It now follows from the hyperconvexity of (X, δ) that there exists $z \in X$ such that

$$\delta(\{z\} \cup \{y_1, \cdots, y_n\}) \leq \sum_{i=1}^{n} r(y_i)$$

for all finite collections $\{y_1, \cdots, y_n\}$ of points of X. In particular this implies $d(z, x) \le r(x)$ for all $x \in X$. Hence $z \in A_i$ for each i and it follows that

$$z \in \bigcap_{i \in I} A_i.$$

It is now immediate that $\cap_{i \in I} A_i \in \mathcal{U}$. It now follows from Zorn's lemma that $\mathcal{U}$ contains an element A which is minimal with respect to set inclusion. We may suppose this element is not a singleton; otherwise it would be a fixed point of f.

Since A is minimal and $f(A) \subseteq A$, it follows that $A = cov(f(A))$; thus

$$A = \bigcap_{x \subset X} B(x; r_x(f(A))).$$

Define

$$d = \sup_{n > 1} \frac{\displaystyle \sup_{x_1, \cdots, x_n \in X} \delta(\{x_1, \cdots, x_n\})}{n}.$$

Since (X, δ) is bounded, there exists $N \in \mathbb{N}$ for which this supremum is attained. Choose $\varepsilon > 0$ so that $\varepsilon \le \frac{d}{N}$. Then there exist $\{y_1, \cdots, y_N\} \subseteq A$ such that

$$\frac{\delta(\{y_1, \cdots, y_N\})}{N} > d - \varepsilon,$$

so it follows that

$$\delta(\{y_1, \cdots, y_N\}) > (N - 1) d.$$

From Property 2 of diversities

$$\sum_{i=2}^{n} d(y_1, y_i) \ge \delta(\{y_1, \cdots, y_N\}) > (N - 1) d.$$

It is now possible to select two points, say x, y, in $\{y_1, \cdots, y_N\}$ such that $d(x, y) > d$.

Now set $A' := A \cap \left(\bigcap_{a \in A} B(a; d) \right)$. We now show that A' is nonempty. First, since A is an admissible set,

$$A = \bigcap_{x \in X} B(x; r_x(A)).$$

Consider the function $\rho : X \to \mathbb{R}$ defined by

$$\rho(x) = \begin{cases} d & \text{if } x \in A \text{ and } r_x(A) > d, \\ r_x(A) & \text{otherwise.} \end{cases}$$

Let $\{y_1, \cdots, y_n\}$ be a finite subset of X, and order these points in such a way that for some $i \in \{0, 1, \cdots, n\}$, $y_i \in A$ and $\rho(y_j) = d$ if $j \leq i$ and $\rho(y_j) = r_{y_j}(A)$ if $j > i$. Then

$$\delta(\{y_1, \cdots, y_i\}) \leq \sup_{x_1, \cdots, x_i \in A} \delta(\{x_1, \cdots, x_i\})$$

$$= i \frac{\sup_{x_1, \cdots, x_i \in A} \delta(\{x_1, \cdots, x_i\})}{i}$$

$$\leq id,$$

and for $j > i + 1$,

$$\rho(y_j) = r_{y_j}(A) \geq d(y_j, y_1) = \delta(\{y_j, y_1\}).$$

Now proceed as follows (if $i = 0$, take y_0 (in place of y_1) to be any point of A)

$$\delta(\{y_1, \cdots, y_n\}) \leq \delta(\{y_1, \cdots, y_i\}) + \sum_{j=i+1}^{n} \delta(\{y_1, y_j\})$$

$$\leq i \cdot d + \sum_{j=i+1}^{n} r_{y_j}(A)$$

$$= \sum_{k=1}^{n} \rho(y_k).$$

Now for $\{x_1, \cdots, x_n\} \subseteq X$, define

$$\rho(\{x_1, \cdots, x_n\}) = \sum_{k=1}^{n} \rho(x_k).$$

By the hyperconvexity of the diversity δ there exists $z \in X$ such that $d(z, a) \leq d$ for any $a \in A$. In particular $z \in \cap_{a \in A} B(a; d)$. Moreover, for any $x \in X$, $d(z, x) \leq \rho(x) \leq r_x(A)$. This implies $z \in A$; hence $z \in A'$. This proves that $A' \neq \emptyset$. Now, since $r_x(A) = r_x(f(A))$ we conclude that A' is f-invariant. On the other hand, $A' \neq A$ because, as shown above, there exist two points, $x, y \in A$ such that $d(x, y) > d$ while $diam(A') \leq d$. This contradiction shows that A must be a singleton consisting of a fixed point for f. $\qquad\square$

QUESTION. We end with a final question. The authors mention in [77] the pattern of the above proof follows the original proof given by Kirk in [110]. As we have seen earlier (see Theorem 3.2) Kirk's theorem also has a constructive proof based on a theorem of Zermelo which avoids an appeal to the Axiom of Choice. This raises the question of whether it is possible to give a more constructive proof Theorem 15.2.

Bibliography

[1] B. Ahmadi Kakavandi, Weak topologies in complete CAT(0) metric spaces. Proc. Am. Math. Soc. **141**(3), 1029–1039 (2013)

[2] B. Ahmadi Kakavandi, M. Amini, Duality and subdifferential for convex functions on complete CAT(0) metric spaces. Nonlinear Anal. **73**(10), 3450–3455 (2010)

[3] A.G. Aksoy, M.S. Borman, A.L. Westfall, Compactness and measures of noncompactness in metric trees, in *Banach and Function Spaces II* (Yokohama Publ., Yokohama, 2008), pp. 277–292

[4] A.G. Aksoy, M.A. Khamsi, *Nonstandard Methods in Fixed Point Theory*. With an introduction by W. A. Kirk. Universitext (Springer, New York, 1990)

[5] A.G. Aksoy, M.A. Khamsi, Fixed points of uniformly lipschitzian mappings in metric trees. Sci. Math. Jpn. **65**(1), 31–41 (2007)

[6] A.G. Aksoy, B. Maurizi, Metric trees, hyperconvex hulls and extensions. Turkish J. Math. **32**(2), 219–234 (2008)

[7] S.M.A. Aleomraninejad, Sh. Rezapour, N. Shahzad, Some fixed point results on a metric space with a graph. Topol. Appl. **159**(3), 659–663 (2012)

[8] A.D. Alexandrov, *Die innere Geometrie der konvexen Flächen* (Akademie-Verlag, Berlin, 1955) (in German)

[9] M.A. Alghamdi, W.A. Kirk, N. Shahzad, Remarks on convex combinations in geodesic spaces. J. Nonlinear Convex Anal. **15**(1), 49–59 (2014)

[10] M.A. Alghamdi, W.A. Kirk, N. Shahzad, Locally nonexpansive mappings in geodesic and length spaces. Topol. Appl. **173**, 59–73 (2014)

[11] D. Ariza-Ruiz, C. Li, G. López-Acedo, The Schauder fixed point theorem in geodesic spaces. J. Math. Anal. Appl. **417**(1), 345–360 (2014)

[12] N. Aronszajn, P. Panitchpakdi, Extension of uniformly continuous transformations and hyperconvex metric spaces. Pacific J. Math. **6**, 405–439 (1956)

[13] H. Aydi, M. Abbas, C. Vetro, Partial Hausdorff metric and Nadler's fixed point theorem on partial metric spaces. Topol. Appl. **159**(14), 3234–3242 (2012)

[14] J.S. Bae, S. Park, Remarks on the Caristi-Kirk fixed point theorem. Bull. Korean Math. Soc. **19**(2), 57–60 (1983)

© Springer International Publishing Switzerland 2014
W. Kirk, N. Shahzad, *Fixed Point Theory in Distance Spaces*,
DOI 10.1007/978-3-319-10927-5

[15] J.S. Bae, Fixed point theorems for weakly contractive multivalued maps. J. Math. Anal. Appl. **284**(2), 690–697 (2003)

[16] J.S. Bae, E.W. Cho, S.H. Yeom, A generalization of the Caristi-Kirk fixed point theorem and its applications to mapping theorems. J. Korean Math. Soc. **31**(1), 29–48 (1994)

[17] J.B. Baillon, Nonexpansive mapping and hyperconvex spaces, in *Fixed Point Theory and Its Applications* (Berkeley, CA, 1986). Contemporary Mathematics, vol. 72 (American Mathematical Society, Providence, 1988), pp. 11–19

[18] I.A. Bakhtin, The contraction mapping principle in almost metric space. Funct. Anal. [Ul'yanovsk. Gos. Ped. Inst., Ul'yanovsk] **30**, 26–37 (1989) (Russian)

[19] I. Bartolini, P. Ciaccia, M. Patella, String matching with metric trees using approximate distance, in *String Processing and Information Retrieval*. Lecture Notes in Computer Science, vol. 2476 (Springer, Berlin, 2002), pp. 271–283

[20] G. Beer, A.L. Dontchev, The weak Ekeland variational principle and fixed points. Nonlinear Anal. **102**, 91–96 (2014)

[21] V.N. Berestovskiĭ, Busemann spaces with upper-bounded Aleksandrov curvature (in Russian). Algebra i Analiz **14**(5), 3–18 (2002) [English transl. in St. Petersburg Math. J. **14**(5), 713–723 (2003)]

[22] V.N. Berestovskiĭ, D.M. Halverson, D. Repovš. Locally G-homogeneous Busemann G-spaces. Differ. Geom. Appl. **29**(3), 299–318 (2011)

[23] V. Berinde, *Contractii Generalizate si Aplicatii*, vol. 22 (Editura Cub Press, Baia Mare, 1997) (in Romanian)

[24] V. Berinde, Generalized contractions in quasimetric spaces, in *Seminar on Fixed Point Theory*, Babes-Bolyai Univ., Cluj-Napoca. Preprint 93-3 (1993), pp. 3–9

[25] M. Bestvina, ℝ-trees in topology, geometry, and group theory, in *Handbook of Geometric Topology* (North-Holland, Amsterdam, 2002), pp. 55–91

[26] I.D. Berg, I.G. Nikolaev, Quasilinearization and curvature of Aleksandrov spaces. Geom. Dedicata **133**, 195–218 (2008)

[27] L.M. Blumenthal, Remarks concerning the Euclidean four-point property. Ergebnisse Math. Kolloq. Wien **7**, 7–10 (1936)

[28] L.M. Blumenthal, *Theory and Applications of Distance Geometry*, 2nd edn. (Chelsea Publishing, New York, 1970)

[29] F. Bojor, Fixed point of ϕ-contraction in metric spaces endowed with a graph. An. Univ. Craiova Ser. Mat. Inform. **37**(4), 85–92 (2010)

[30] J.M. Borwein, Completeness and the contraction principle. Proc. Am. Math. Soc. **87**(2), 246–250 (1983)

[31] M. Bota, A. Molnár, C. Varga, On Ekeland's variational principle in b-metric spaces. Fixed Point Theory **12**(2), 21–28 (2011)

[32] A. Bottaro Aruffo, G. Bottaro, Some variational results using generalizations of sequential lower semicontinuity. Fixed Point Theory Appl. **2010**, 21 pp. (2010). Art. ID 323487

[33] S. Bouamama, D. Misane, Hyperconvex ultrametric spaces and fixed point theory. New Zealand J. Math. **34**(1), 25–29 (2005)

[34] H. Brézis, F.E. Browder, A general principle on ordered sets in nonlinear functional analysis. Adv. Math. **21**(3), 355–364 (1976)

[35] A. Branciari, A fixed point theorem of Banach-Caccioppoli type on a class of generalized metric spaces. Publ. Math. Debrecen **57**(1–2), 31–37 (2000)

[36] M. Bridson, A. Haefliger, *Metric Spaces of Non-positive Curvature* (Springer, Berlin/Heidelberg/New York, 1999)

[37] N. Brodskiy, J. Dydak, J. Higes, A. Mitra, Dimension zero at all scales. Topol. Appl. **154**(14), 2729–2740 (2007)

[38] A. Brøndsted, Fixed points and partial orders. Proc. Am. Math. Soc. **60**, 365–366 (1976)

[39] F.E. Browder, On a theorem of Caristi and Kirk, in *Fixed Point Theory and Its Applications* (Proc. Sem., Dalhousie Univ., Halifax, NS, 1975) (Academic, New York, 1976), pp. 23–27

[40] F.E. Browder, Semicontractive and semiaccretive nonlinear mappings in a Banach space. Bull. Am. Math. Soc. **74**, 660–665 (1968)

[41] F. Bruhat, J. Tits, Groupes réductifs sur un corps local. I. Données radicielles valuées. Inst. Hautes Études Sci. Publ. Math. **41**, 5–251 (1972) (in French)

[42] N. Brunner, Topologische maximalprinzipien. Math. Logik Grundlag. Math. **33**(2), 135–139 (1987) (in German)

[43] D. Bryant, P.F. Tupper, Hyperconvexity and tight-span theory for diversities. Adv. Math. **231**(6), 3172–3198 (2012)

[44] S.M. Buckley, K. Falk, D.J. Wraith, Ptolemaic spaces and CAT(0). Glasgow Math. J. **51**(2), 301–314 (2009)

[45] D. Burago, Y. Burago, S. Ivanov, *A Course in Metric Geometry*. Graduate Studies in Mathematics, vol. 33 (American Mathematical Society, Providence, 2001)

[46] H. Busemann, *The Geometry of Geodesics* (Academic, New York, 1955)

[47] S.V. Buyalo, Geodesics in Hadamard Spaces. Algebra i Analiz **10**(2), 93–123 (1998) (Russian); translation in St. Petersburg Math. J. **10**(2), 293–313 (1999)

[48] A. Całka, On conditions under which isometries have bounded orbits. Colloq. Math. **48**(2), 219–227 (1984)

[49] J. Caristi, Fixed point theorems for mappings satisfying inwardness conditions. Trans. Am. Math. Soc. **215**, 241–251 (1976)

[50] J. Caristi, Fixed point theory and inwardness conditions, in *Applied Nonlinear Analysis* (Proceedings of Third International Conference, University of Texas, Arlington, TX, 1978) (Academic, New York/London, 1979), pp. 479–483

[51] E. Cartan, *Leçons sur la Géometrié des Espaces de Riemann*, 2nd edn. (Gauthier-Villars, Paris, 1951) (in French)

[52] Y. Chen, Y.J. Cho, L. Yang, Note on the results with lower semi-continuity. Bull. Korean Math. Soc. **39**(4), 535–541 (2002)

[53] E.W. Chittenden, On the equivalence of écart and voisinage. Trans. Am. Math. Soc. **18**(2), 161–166 (1917)

[54] P. Corazza, Introduction to metric preserving functions. Am. Math. Mon. **106**(4), 309–323 (1999)

[55] G. Cortelazzo, G. Mian, G. Vezzi, P. Zamperoni, Trademark shapes description by string matching techniques. Pattern Recognit. **27**(8), 1005–1018 (1994)

[56] M. Crandall, A. Pazy, Semi-groups of nonlinear contractions and dissipative sets. J. Funct. Anal. **3**, 376–418 (1969)

[57] S. Czerwik, Nonlinear set-valued contraction mappings in b-metric spaces. Atti Sem. Mat. Fis. Univ. Modena **46**(2), 263–276 (1998)

[58] S. Czerwik, Contraction mappings in b-metric spaces. Acta Math. Inform. Univ. Ostraviensis **1**, 5–11 (1993)

[59] H. Dehghan, J. Rooin, Metric projection and convergence theorems for nonexpansive mappings in Hadamard spaces. arXiv:1410.1137[math.FA] (2014)

[60] S. Dhompongsa, W.A. Kirk, B. Sims, Fixed points of uniformly lipschitzian mappings. Nonlinear Anal. **65**(4), 762–772 (2006)

[61] S. Dhomponga, W.A. Kirk, B. Panyanak, Nonexpansive set-valued mappings in metric and Banach spaces. J. Nonlinear Convex Anal. **8**(1), 35–45 (2007)

[62] A.L. Dontchev, W.W. Hager, An inverse mapping theorem for set-valued maps. Proc. Am. Math. Soc. **121**(2), 481–489 (1994)

[63] D. Downing, W.A. Kirk, A generalization of Caristi's theorem with applications to nonlinear mapping theory. Pacific J. Math. **69**(2), 339–346 (1977)

[64] A. Dress, Trees, tight extensions of metric trees, and the cohomological dimension of certain groups: a note on combinatorial properties of metric spaces. Adv. Math. **53**(3), 321–402 (1984)

[65] A. Dress, R. Scharlau, Gated sets in metric spaces. Aequationes Math. **34**(1), 112–120 (1987)

[66] A. Dress, W.F. Terhalle, The real tree. Adv. Math. **120**(2), 283–301 (1996)

[67] N. Dunford, J.T. Schwartz, *Linear Operators, Part I: General Theory* (Wiley Interscience, New York, 1957)

[68] M. Edelstein, On non-expansive mappings of Banach spaces. Proc. Camb. Philos. Soc. **60**, 439–447 (1964)

[69] R. Engelking, *Theory of Dimensions Finite and Infinite* (Heldermann Verlag, Lemgo, 1995)

[70] I. Ekeland, Sur les problèmes variationnels. C. R. Acad. Sci. Paris Sér. A-B **275**, A1057–A1059 (1972)

[71] I. Ekeland, On the variational principle. J. Math. Anal. Appl. **47**, 324–353 (1974)

[72] R. Espínola, M.A. Khamsi, Introduction to hyperconvex spaces, in *Handbook of Metric Fixed Point Theory* (Kluwer Academic Publishers, Dordrecht, 2001), pp. 391–435.

[73] R. Espínola, W.A. Kirk, Fixed point theorems in $\mathbb{R}$-trees with applications to graph theory. Topol. Appl. **153**(7), 1046–1055 (2006)

[74] R. Espínola, A. Fernández-León, CAT(κ) spaces, weak convergence and fixed points. J. Math. Anal. Appl. **353**(1), 410–427 (2009)

[75] R. Espínola, A. Nicolae, Geodesic Ptolemy spaces and fixed points. Nonlinear Anal. **74**(1), 27–34 (2011)

[76] R. Espínola, B. Piątek, The fixed point property and unbounded sets in CAT(0) spaces. J. Math. Anal. Appl. **408**(2), 638–654 (2013)

[77] R. Espínola, B. Piątek, Diversities, hyperconvexity and fixed points. Nonlinear Anal. **95**, 229–245 (2014)

[78] R. Fagin, R. Kumar, D. Sivakumar, Comparing top k lists. SIAM J. Discrete Math. **17**(1), 134–160 (2003)

[79] R. Fagin, L. Stockmeyer, Relaxing the triangle inequality in pattern matching. Int. J. Comput. Vis. **30**(3), 219–231 (1998)

[80] D.P. Faith, Conservation evaluation and phylogenetic diversity. Biol. Conserv. **61**, 1–10 (1992)

[81] K. Fan, Extensions of two fixed point theorems of F. E. Browder. Math. Z. **112**, 234–240 (1969)

[82] T. Foertsch, A. Lytchak, V. Schroeder, Nonpositive curvature and the Ptolemy inequality. Int. Math. Res. Not. **2007**(22), 15 pp. (2007). Article ID rnm100. Erratum to: "Nonpositive curvature and the Ptolemy inequality". Int. Math. Res. Not. IMRN **2007**(24), 1 pp. Art. ID rnm160

[83] T. Foertsch, V. Schroeder, Group actions on geodesic Ptolemy spaces. Trans. Am. Math. Soc. **363**(6), 2891–2906 (2011)

[84] M. Frigon, On continuation methods for contractive and nonexpansive mappings, in *Recent Advances on Metric Fixed Point Theory* (Seville, 1995), Ciencias, vol. 48 (University of Sevilla, Seville, 1996), pp. 19–30

[85] B. Fuchssteiner, Iterations and fixpoints. Pacific J. Math. **68**(1), 73–80 (1977)

[86] K. Goebel, W.A. Kirk, Iteration processes for nonexpansive mappings, in *Topological methods in nonlinear functional analysis (Toronto, Ont., 1982)*, Contemporary Mathematics, vol. 21 (American Mathematical Society, Providence, 1983), pp. 115–123

[87] K. Goebel, W.A. Kirk, *Topics in Metric Fixed Point Theory*. Cambridge Studies in Advanced Mathematics, vol. 28 (Cambridge University Press, Cambridge, 1990)

[88] K. Goebel, S. Reich, *Uniform Convexity, Hyperbolic Geometry, and Nonexpansive Mappings* (Marcel Dekker, New York/Basel, 1984)

[89] A. Granas, Continuation methods for contractive maps. Topol. Methods Nonlinear Anal. **3**(2), 375–379 (1994)

[90] R.H. Haghi, Sh. Rezapour, N. Shahzad, Be careful on partial metric fixed point results. Topol. Appl. **160**(3), 450–454 (2013)

[91] B. Halpern, Fixed points of nonexpanding maps. Bull. Am. Math. Soc. **73**, 957–961 (1967)

[92] J. Heinonen, *Lectures on Analysis on Metric Spaces*. Universitext (Springer, New York, 2001)

[93] S.K. Hildebrand, H.W. Milnes, Minimal arcs in metric spaces. J. Austral. Math. Soc. **19**(4), 426–430 (1975)

[94] P. Hitzelberger, A. Lytchak, Spaces with many affine functions. Proc. Am. Math. Soc. **135**(7), 2263–2271 (2007)

[95] P. Hitzler, A. Seda, *Mathematical Aspects of Logic Programming Semantics* (Chapman & Hall/CRC Studies in Informatics Series, CRC Press, 2010)

[96] R.D. Holmes, Fixed points for local radial contractions (Proc. Sem., Dalhousie Univ., Halifax, NS, 1975). (Academic, New York, 1976), pp. 79–89

[97] T. Hu, W.A. Kirk, Local contractions in metric spaces. Proc. Am. Math. Soc. **68**(1), 121–124 (1978)

[98] P.N. Ivanshin, Properties of two selections in metric spaces of nonpositive curvature. Asian-Eur. J. Math. **1**(3), 383–395 (2008)

[99] J.R. Isbell, Six theorems about injective metric spaces. Comment. Math. Helv. **39**, 65–76 (1964)

[100] J. Jachymski, The contraction principle for mappings on a metric space with a graph. Proc. Am. Math. Soc. **136**(4), 1359–1373 (2008)

[101] J. Jachymski, J. Matkowski, T. Świątkowski, Nonlinear contractions on semimetric spaces. J. Appl. Anal. **1**(2), 125–134 (1995)

[102] M. Jleli, B. Samet, The Kannan's fixed point theorem in a cone rectangular metric space. J. Nonlinear Sci. Appl. **2**(3), 161–167 (2009)

[103] G. Jungck, Local radial contractions – a counter-example. Houston J. Math. **8**(4), 501–506 (1982)

[104] Z. Kadelburg, S. Radenović, On generalized metric spaces: a survey. TWMS J. Pure Appl. Math. **5**(1), 3–13 (2014)

[105] M.A. Khamsi, On asymptotically nonexpansive mappings in hyperconvex metric spaces. Proc. Am. Math. Soc. **132**(2), 365–373 (2004)

[106] M.A. Khamsi, Remarks on Caristi's fixed point theorem. Nonlinear Anal. **71**(1–2), 227–231 (2009)

[107] M.A. Khamsi, W.A. Kirk, *An Introduction to Metric Spaces and Fixed Point Theory*. Pure and Applied Mathematics (Wiley-Interscience, New York, 2001)

[108] W.A. Kirk, On locally isometric mappings of a G-space on itself. Proc. Am. Math. Soc. **15**, 584–586 (1964)

[109] W.A. Kirk, Isometries in G-spaces. Duke Math. J. **31**, 539–541 (1964)

[110] W.A. Kirk, A fixed point theorem for mappings which do not increase distances. Am. Math. Mon. **72**, 1004–1006 (1965)

[111] W.A. Kirk, A theorem on local isometries. Proc. Am. Math. Soc. **17**, 453–455 (1966)

[112] W.A. Kirk, On conditions under which local isometries are motions. Colloq. Math. **22**, 229–232 (1971)

[113] W.A. Kirk, Caristi's fixed point theorem and metric convexity. Colloq. Math. **36**(1), 91–86 (1976)

[114] W.A. Kirk, An abstract fixed point theorem for nonexpansive mappings. Proc. Am. Math. Soc. **82**, 640–642 (1981)

[115] W.A. Kirk, History and methods of metric fixed point theory, in *Antipodal Points and Fixed Points*. Lecture Notes Series, vol. 28 (Seoul National University, Seoul, 1995), pp. 21–54

[116] W.A. Kirk, Hyperconvexity of $\mathbb{R}$-trees. Fund. Math. **156**(1), 67–72 (1998)

[117] W.A. Kirk, Contraction mappings and extensions, in *Handbook of Metric Fixed Point Theory* (Kluwer Academic Publishers, Dordrecht, 2001), pp. 1–34

[118] W.A. Kirk, Geodesic geometry and fixed point theory, in *Seminar of Mathematical Analysis* (Malaga/Seville, 2002/2003). Colecc. Abierta, vol. 64 (Univ. Sevilla Secr. Publ., Seville, 2003), pp. 195–225

[119] W.A. Kirk, Geodesic geometry and fixed point theory. II, in *International Conference on Fixed Point Theory and Applications* (Yokohama Publ., Yokohama, 2004), pp. 113–142

[120] W.A. Kirk, Fixed point theorems in CAT(0) spaces and $\mathbb{R}$-trees. Fixed Point Theory Appl. **2004**(4), 309–316 (2004)

[121] W.A. Kirk, Approximate fixed points of nonexpansive maps. Fixed Point Theory **10**(2), 275–288 (2009)

[122] W.A. Kirk, Remarks on approximate fixed points. Nonlinear Anal. **75**(12), 4632–4636 (2012)

[123] W.A. Kirk, B. Panyanak, Best approximation in $\mathbb{R}$-trees. Numer. Funct. Anal. Optim. **28**(5–6), 681–690 (2007)

[124] W.A. Kirk, B. Panyanak, A concept of convergence in geodesic spaces. Nonlinear Anal. **68**(12), 3689–3696 (2008)

[125] W.A. Kirk, S. Massa, Remarks on asymptotic and Chebyshev centers. Houston J. Math. **16**(3), 357–363 (1990)

[126] W.A. Kirk, L.M. Saliga, The Brézis-Browder order principle and extensions of Caristi's theorem, Proceedings of the Third World Congress of Nonlinear Analysts, Part 4 (Catania, 2000). Nonlinear Anal. **47**(4), 2765–2778 (2001)

[127] W.A. Kirk, N. Shahzad, Some fixed point results in ultrametric spaces. Topol. Appl. **159**(15), 3327–3334 (2012)

[128] W.A. Kirk, N. Shahzad, Remarks on metric transforms and fixed-point theorems. Fixed Point Theory Appl. **2013**, 11 pp. (2013)

[129] W.A. Kirk, N. Shahzad, Generalized metrics and Caristi's theorem. Fixed Point Theory Appl. **2014**, 3 pp. (2014)

[130] W.A. Kirk, N. Shahzad, Uniformly lipschitzian mappings in $\mathbb{R}$-trees. J. Nonlinear Convex Anal. (to appear)

[131] W.A. Kirk, B. Sims, An ultrafilter approach to locally almost nonexpansive maps. Nonlinear Anal. **63**, e1241–e1251 (2005)

[132] B. Kleiner, Lectures on spaces of nonpositive curvature by W. Ballmann; Metric spaces of non-positive curvature by Martin R. Bridson and André Haefliger; Geometry of nonpositively curved manifolds by Patrick B. Eberlein. Bull. Am. Math. Soc. **39**, 273–279 (2002, in Review)

[133] U. Kohlenbach, Some logical metatheorems with applications in functional analysis. Trans. Am. Math. Soc. **357**(1), 89–128 (2005)

[134] B. Krakus, Any 3-dimensional G-space is a manifold. Bull. Acad. Polon. Sci. Sér. Sci. Math. Astronom. Phys. **16**, 737–740 (1968)

[135] T. Kuczumow, An almost convergence and its applications. Ann. Univ. Mariae Curie-Skłodowska Sect. A **32**, 79–88 (1978)

[136] J. Kulesza, T.C. Lim, On weak compactness abd countable weak compactness in fixed point theory. Proc. Am. Math. Soc. **124**, 3345–3349 (1996)

[137] U. Lang, V. Schroeder, Jung's theorem for Alexandrov spaces of curvature bounded above. Ann. Glob. Anal. Geom. **15**(3), 263–275 (1997)

[138] A. Lemin, On ultrametrization of general metric spaces. Proc. Am. Math. Soc. **131**(3), 979–989 (2003)

[139] Z. Li, Remarks on Caristi's fixed point theorem and Kirk's problem. Nonlinear Anal. **73**(12), 3751–3755 (2010)

[140] E.A. Lifšic, A fixed point theorem for operators in strongly convex spaces. Voronež. Gos. Univ. Trudy Mat. Fak. Vyp. (Sb. Stateĭ po Nelineĭnym Operator. Uravn. i Priložen) **16**, 23–28 (1975, Russian)

[141] T.C. Lim, Remarks on some fixed point theorems. Proc. Am. Math. Soc. **60**, 179–182 (1976)

[142] B. Lins, Asymptotic behavior of nonexpansive mappings in finite dimensional normed spaces. Proc. Am. Math. Soc. **137**(7), 2387–2392 (2009)

[143] R. Mańka, Some forms of the axiom of choice. Jbuch. Kurt-Gödel-Ges., Wien **1**, 24–34 (1988)

[144] R. Mańka, On generalized methods of successive approximations. Non-linear Anal. **72**(3–4), 1438–1444 (2010)

[145] J.T. Markin, Fixed points, selections and best approximation for multivalued mappings in ℝ-trees. Nonlinear Anal. **67**(9), 2712–2716 (2007)

[146] J. Markin, N. Shahzad, Fixed point theorems for inward mappings in ℝ-trees. J. Nonlinear Convex Anal. **16** (2015) (in press)

[147] J. Martínez-Maurica, M.T. Pellón, Non-archimedean Chebyshev centers. Nederl. Akad. Wetensch. Indag. Math. **49**(4), 417–421 (1987)

[148] J. Matkowski, Integrable solutions of functional equations. Dissertationes Math. (Rozprawy Mat.) **127**, 68 pp. (1975)

[149] S.G. Matthews, Partial metric topology, in *Papers on General Topology and Applications* (Flushing, NY, 1992). Annals of the New York Academy of Sciences, vol. 728 (New York Academy of Sciences, New York, 1994), pp. 183–197

[150] S.G. Matthews, An extensional treatment of lazy data flow deadlock. Topology and completion in semantics (Chartres, 1993). Theor. Comput. Sci. **151**(1), 195–205 (1995)

[151] J.C. Mayer, L.G. Oversteegen, A topological characterization of ℝ-trees. Trans. Am. Math. Soc. **320**(1), 395–415 (1990)

[152] R. McConnell, R. Kwok, J. Curlander, W. Kober, S. Pang, Ψ-S correlation and dynamic time warping: two methods for tracking ice floes. IEEE Trans. Geosci. Remote Sens. **29**(6), 1004–1012 (1991)

[153] K. Menger, Untersuchungen über allgemeine Metrik. Math. Ann. **100**(1), 75–163 (1928) (in German)

[154] P.R. Meyers, A converse to Banach's contraction theorem. J. Res. Nat. Bur. Stand. Sect. B **71B**, 73–76 (1967)

[155] B. Minh, S. Klaere, A. von Haeseler, Taxon selection under split diversity. Syst. Biol. **58**(6), 586–594 (2009)

[156] A.F. Monna, *Analyse Non-archimedienne* (Springer, Berlin/ Hiedelberg/New York, 1970) (in French)

[157] N. Monod, Supperrigidity for irreducible lattices and geometric splitting. J. Am. Math. Soc. **19**(4), 781–814 (2006)

[158] J. Mycielski, On the existence of a shortest arc between two points of a metric space. Houston J. Math. **20**(3), 491–494 (1994)

[159] S.B. Nadler Jr., Multivalued contraction mappings. Pacific J. Math. **30**, 475–488 (1969)

[160] L. Narici, E. Beckenstein, G. Bachman, *Functional Analysis and Valuation Theory* (Marcel Dekker, New York, 1971)

[161] R. Nowakowski, I. Rival, Fixed-edge theorem for graphs with loops. J. Graph Theory **3**(4), 339–350 (1979)

[162] R.D. Nussbaum, Degree theory for local condensing mappings. J. Math. Anal. Appl. **37**, 741–766 (1972)

[163] S. Oltra, S. Romaguera, E.A. Sánchez-Pérez, Bicompleting weightable quasi-metric spaces and partial metric spaces. Rend. Circ. Mat. Palermo (2) **51**(1), 151–162 (2002)

[164] S. Oltra, O. Valero, Banach's fixed point theorem for partial metric spaces. Rend. Istid. Math. Univ. Trieste **36**(1–2), 17–26 (2004)

[165] M.A. Ostrowski, The round-off stability of iterations. Z. Angew. Math. Mech. **47**, 77–81 (1967)

[166] A. Papadopoulos, *Metric Spaces, Convexity and Nonpositive Curvature*. IRMA Lectures in Mathematics and Theoretical Physics, vol. 6 (European Mathematical Society (EMS), Zürich, 2005)

[167] L. Pasicki, A short proof of the Caristi theorem. Comment. Math. Prace Mat. **20**(2), 427–428 (1977/1978)

[168] L. Pachter, D. Speyer, Reconstructing trees from subtree weights. Appl. Math. Lett. **17**(6), 615–621 (2004)

[169] J.-P. Penot, A short constructive proof of Caristi's fixed piont theorem. Publ. Math. Univ. Paris **10**, 1–3 (1976)

[170] C. Petalas, T. Vidalis, A fixed point theorem in non-archimedean vector spaces. Proc. Am. Math. Soc. **118**(3), 819–821 (1993)

[171] A. Petrusel, I.A. Rus, Fixed point theorems in ordered L-spaces. Proc. Am. Math. Soc. **134**(2), 411–418 (2006)

[172] P. Pongsriiam, I. Termwuttipong, On metric preserving functions and fixed point theorems. Fixed Point Theory Appl. **2014**, 14 pp. (2014)

[173] R. Precup, Continuation results for mappings of contractive type. Semin. Fixed Point Theory Cluj-Napoca **2**, 23–40 (2001)

[174] S. Priess-Crampe, Der Banachsche Fixpunktsatz für ultrametrische Räume. Results Math. **18**(1–2), 178–186 (1990) (in German)

[175] S. Priess-Crampe, Some results of functional analysis for ultrametric spaces and valued vector spaces. Geom. Dedicata **58**(1), 79–90 (1995)

[176] S. Priess-Crampe, Remarks on some theorems of functional analysis, in *Ultrametric Functional Analysis*. Contemporary Mathematics, vol. 384 (American Mathematical Society, Providence, 2005), pp. 235–246

[177] S. Priess-Crampe, P. Ribenboim, Generalized ultrametric spaces II. Abh. Math. Sem. Univ. Hamburg **67**, 19–31 (1997)

[178] S. Priess-Crampe, P. Ribenboim, The common point theorem for ultrametric spaces. Geom. Dedicata **72**(1), 105–110 (1998)

[179] S. Priess-Crampe, P. Ribenboim, Fixed point and attractor theorems for ultrametric spaces. Forum Math. **12**(1), 53–64 (2000)

[180] S. Priess-Crampe, P. Ribenboim, Ultrametric dynamics. Illinois J. Math. **55**(1), 287–303 (2011)

[181] S. Priess-Crampe, P. Ribenboim, The approximation to a fixed point. J. Fixed Point Theory Appl. **14**(1), 41–53 (2013)

[182] E. Rakotch, A note on α-locally contractive mappings. Bull. Res. Council Israel Sect. F **10F**, 188–191 (1962)

[183] W.O. Ray, The fixed point property and unbounded sets in Hilbert space. Trans. Am. Math. Soc. **258**(2), 531–537 (1980)

[184] S. Reich, Strong convergence theorems for resolvents of accretive operators in Banach spaces. J. Math. Anal. Appl. **75**(1), 287–292 (1980)

[185] S. Reich, The almost fixed point property for nonexpansive mappings. Proc. Am. Math. Soc. **88**(1), 44–46 (1983)

[186] S. Reich, I. Shafrir, Nonexpansive iterations in hyperbolic spaces. Nonlinear Anal. **15**(6), 537–558 (1990)

[187] J. Reinermann, R. Schöneberg, Some results and problems in the fixed point theory for nonexpansive and pseudocontractive mappings in Hilbert space, in *Fixed Point Theory and Its Applications* (Proc. Sem., Dalhousie Univ., Halifax, NS, 1975) (Academic, New York, 1976), pp. 187–196

[188] D. Repovš, *Mathematical Reviews* (American Mathematical Society, Providence, 1997) [97m:57030]

[189] S. Romaguera, A Kirk type characterization of completeness for partial metric spaces. Fixed Point Theory Appl. **2010**, 6 pp. (2010). Article ID 493298

[190] S. Romaguera, On Nadler's fixed point theorem for partial metric spaces. Math. Sci. Appl. E-Notes **1**, 7 pp. (2013)

[191] I.A. Rus, *Generalized Contractions and Applications* (Cluj University Press, Cluj-Napoca, 2001)

[192] M. Samreen, T. Kamran, Fixed point theorems for integral G-contractions. Fixed Point Theory Appl. **2013**, 11 pp. (2013)

[193] M. Samreen, T. Kamran, N. Shahzad, Some fixed point theorems in b-metric spaces endowed with graph. Abstr. Appl. Anal. **2013**, 9 pp. (2013). Article ID 967132

[194] I.R. Sarma, J.M. Rao, S.S. Rao, Contractions over generalized metric spaces. J. Nonlinear Sci. Appl. **2**(3), 180–182 (2009)

[195] C. Semple, *Phylogenetics*. Oxford Lecture Series in Mathematics and Its Applications (Oxford University Press, Oxford, 2003)

[196] I.J. Schoenberg, A remark on M. M. Day's characterization of inner-product spaces and a conjecture of L. M. Blumenthal. Proc. Am. Math. Soc. **3**, 961–964 (1952)

[197] I. Shafrir, The approximate fixed point property in Banach and hyperbolic spaces. Israel J. Math. **71**(2), 211–223 (1990)

[198] N. Shahzad, Fixed point results for multimaps in CAT(0) spaces. Topol. Appl. **156**(5), 997–1001 (2009)

[199] N. Shahzad, J. Markin, Invariant approximations for commuting mappings in CAT(0) and hyperconvex spaces. J. Math. Anal. Appl. **337**(2), 1457–1464 (2008)

[200] N. Shahzad, O. Valero, On 0-complete partial metric spaces and quantitative fixed point techniques. Abstr. Appl. Anal. **2013**, 11 pp. (2013). Art. ID 985095

[201] S. Shirali, Maps for which some power is a contraction. Math. Commun. **15**(1), 139–141 (2010)

[202] J. Siegel, A new proof of Caristi's fixed point theorem. Proc. Am. Math. Soc. **66**(1), 54–56 (1977)

[203] B. Sims, H.K. Xu, Locally almost nonexpansive mappings. Commun. Appl. Nonlinear Anal. **8**(3), 81–88 (2001)

[204] S.P. Singh, B. Watson, P. Srivastava, *Fixed Point Theory and Best Approximation: The KKM-Map Principle* (Kluwer, Dordrecht, 1997)

[205] M.A. Steel, Phylogenetic diversity and the Greedy algorithm. Syst. Biol. **54**(4), 527–529 (2005)

[206] T. Suzuki, Generalized Caristi's fixed point theorems by Bae and others. J. Math. Anal. Appl. **302**(2), 502–508 (2005)

[207] T. Suzuki, A generalized Banach contraction principle that characterizes metric completeness. Proc. Am. Math. Soc. **136**(5), 1861–1869 (2008)

[208] W. Takahashi, A convexity in metric space and nonexpansive mappings. I. Kōdai Math. Sem. Rep. **22**, 142–149 (1970)

[209] P. Thurston, 4-Dimensional Busemann G-spaces are 4-manifolds. Differ. Geom. Appl. **6**(3), 245–270 (1996)

[210] J. Tits, A "theorem of Lie-Kolchin" for trees, in *Contributions to Algebra* (collection of papers dedicated to Ellis Kolchin) (Academic, New York, 1977), pp. 377–388

[211] M. Turinici, Functional contractions in local Branciari metric spaces. ROMAI J. **8**(2), 189–199 (2012)

[212] M. van de Vel, *Theory of Convex Structures* (North Holland, Amsterdam, 1993)

[213] A.C.M. van Rooij, *Non-archimedean Functional Analysis* (Marcel Dekker, New York/Basel, 1978)

[214] L.E. Ward Jr., Recent developments in dendritic spaces and related topics, in *Studies in Topology* (Proc. Conf., Univ. North Carolina, Charlotte, NC, 1974; dedicated to Math. Sect. Polish Acad. Sci.) (Academic, New York, 1975), pp. 601–647

[215] J.D. Weston, A characterization of metric completeness. Proc. Am. Math. Soc. **64**(1), 186–188 (1977)

[216] W.A. Wilson, On semi-metric spaces. Am. J. Math. **53**(2), 361–373 (1931)

[217] C.S. Wong, On a fixed point theorem of contractive type. Proc. Am. Math. Soc. **57**(2), 283–284 (1976)

[218] Q. Xia, The geodesic problem in quasimetric spaces. J. Geom. Anal. **19**(2), 452–479 (2009)

[219] G.S. Young, The introduction of local connectivity by change of topology. Am. J. Math. **68**, 479–494 (1946)

[220] G.S. Young, Fixed-point theorems for arcwise connected continua. Proc. Am. Math. Soc. **11**, 880–884 (1960)

[221] E. Zeidler, *Nonlinear Functional Analysis and its Applications I: Fixed Point Theorems* (Springer, Berlin, 1986)

[222] E. Zermelo, Neuer Beweis für die Möglichkeit einer Wohlordnung. Math. Ann. **65**(1), 107–128 (1907) (in German)

[223] G. Zhang, D. Jiang, On the fixed point theorems of Caristi type. Fixed Point Theory **14**(2), 523–529 (2013)

Index

© Springer International Publishing Switzerland 2014
W. Kirk, N. Shahzad, *Fixed Point Theory in Distance Spaces*,
DOI 10.1007/978-3-319-10927-5

Printed in the United States
By Bookmasters